Atmospheric Conditions of Forestry

Atmospheric Conditions of Forestry

Pawan Pattanaik

RANDOM

RANDOM PUBLICATIONS

NEW DELHI (INDIA)

Atmospheric Conditions of Forestry

ISBN 978-93-5111-608-0

Published in 2015 in India by

Reprint 2019

RANDOM PUBLICATIONS

4376-A/4B, Gali Murari Lal, Ansari Road
New Delhi-110 002
Phone : +9111-43580356, 011-23289044, 011-43142548
e-mail: sales@randompublications.com,
info@randompublications.com, randomexports@gmail.com

Type Setting by : Friends Media, Delhi-110089
Printed at : Mehra Printers, Delhi-110 092

Preface

Forestry in India is a significant rural industry and a major environmental resource. India is one of the ten most forest-rich countries of the world along with the Russian Federation, Brazil, Canada, United States of America, China, Democratic Republic of the Congo, Australia, Indonesia and Sudan. Together, India and these countries account for 67 percent of total forest area of the world. India's forest cover grew at 0.22% annually over 1990-2000 and has grown at the rate of 0.46% per year over 2000-2010, after decades where forest degradation was a matter of serious concern. Climate changes directly and indirectly affect the growth and productivity of forests: directly due to changes in atmospheric carbon dioxide and climate and indirectly through complex interactions in forest ecosystems. Climate also affects the frequency and severity of many forest disturbances. In conjunction with the projected impacts of climate change, forests face impacts from land development, suppression of natural periodic forest fires, and air pollution. Although it is difficult to separate the effects of these different factors, the combined impact is already leading to changes in our forests. As these conditions are likely to continue in the decades ahead, some of the valuable goods and services provided by forests may be compromised.

I would like to thank my team for standing beside me throughout my career and writing this book. My special thanks go to "Random Publications" who have published the book.

– Pawan Pattanaik

Contents

1

Atmosphere and Forest-An Introduction

ATMOSPHERIC COMPOSITION

The table indicates that nitrogen and oxygen are the main components of the atmosphere by volume. Together these two gases make up approximately 99% of the dry atmosphere. Both of these gases have very important associations with life. Nitrogen is removed from the atmosphere and deposited at the Earth's surface mainly by specialized nitrogen fixing bacteria, and by way of lightning through precipitation. The addition of this nitrogen to the Earth's surface soils and various water bodies supplies much needed nutrition for plant growth. Nitrogen returns to the atmosphere primarily through biomass combustion and denitrification.

Table. Average composition of the atmosphere up to an altitude of 25 km.

Gas Name	Chemical Formula	Percent Volume
Nitrogen	N2	78.08%
Oxygen	O2	20.95%
Water	H2O	0 to 4%
Argon	Ar	0.93%
Carbon Dioxide	CO2	0.0360%
Neon	Ne	0.0018%
Helium	He	0.0005%
Methane	CH4	0.00017%
Hydrogen	H2	0.00005%
Nitrous Oxide	N2O	0.00003%
Ozone	O3	0.000004%

Oxygen is exchanged between the atmosphere and life through the processes of photosynthesis and respiration. Photosynthesis produces oxygen when carbon dioxide and water are chemically converted into glucose with the help of sunlight. Respiration is a the opposite process of photosynthesis. In respiration, oxygen is combined with glucose to chemically release energy for metabolism. The products of this reaction are water and carbon dioxide.

The next most abundant gas on the table is water vapor. Water vapor varies in concentration in the atmosphere both spatially and temporally. The highest concentrations of water vapor are found near the equator over the oceans and tropical rain forests. Cold polar areas and subtropical continental deserts are locations where the volume of water vapor can approach zero percent. Water vapor has several very important functional roles on our planet:

- It redistributes heat energy on the Earth through latent heat energy exchange.
- The condensation of water vapor creates precipitaion that falls to the Earth's surface providing needed fresh water for plants and animals.
- It helps warm the Earth's atmosphere through the greenhouse effect.

The fifth most abundant gas in the atmosphere is carbon dioxide. The volume of this gas has increased by over 35% in the last three hundred years. This increase is primarily due to human induced burning from fossil fuels, deforestation, and other forms of land-use change. Carbon dioxide is an important greenhouse gas.

The human-caused increase in its concentration in the atmosphere has strengthened the greenhouse effect and has definitely contributed to global warming over the last 100 years. Carbon dioxide is also naturally exchanged between the atmosphere and life through the processes of photosynthesis and respiration.

Methane is a very strong greenhouse gas. Since 1750, methane concentrations in the atmosphere have increased by more than 150%. The primary sources for the additional methane added to the atmosphere are: rice cultivation; domestic grazing animals; termites; landfills; coal mining; and, oil and gas extraction. Anaerobic conditions associated with rice paddy flooding results in the formation of methane gas. However, an accurate estimate of how much methane is being produced from rice paddies has been difficult to ascertain.

More than 60% of all rice paddies are found in India and China where scientific data concerning emission rates are unavailable. Nevertheless, scientists believe that the contribution of rice paddies is large because this form of crop production has more than doubled since 1950. Grazing animals release methane to the environment as a result of herbaceous digestion. Some researchers believe the addition of methane from this source has more than quadrupled over the last century.

Termites also release methane through similar processes. Land-use change in the tropics, due to deforestation, ranching, and farming, may be causing termite numbers to expand. If this assumption is correct, the contribution from these insects may be important. Methane is also released from landfills, coal mines, and gas and oil drilling. Landfills produce methane as organic wastes decompose over time. Coal, oil, and natural gas deposits release methane to the atmosphere when these deposits are excavated or drilled. The average concentration of the greenhouse gas nitrous oxide is now increasing at a rate of 0.2 to 0.3% per year. Its part in the enhancement of the greenhouse effect is minor relative to the other greenhouse gases already mentioned. However, it does have an important role in the artificial fertilization of ecosystems. In extreme cases, this fertilization can lead to the death of forests, eutrophication of aquatic habitats, and species exclusion. Sources for the increase of nitrous oxide in the atmosphere include: land-use conversion; fossil fuel combustion; biomass burning; and soil fertilization.

Most of the nitrous oxide added to the atmosphere each year comes from deforestation and the conversion of forest, savanna and grassland ecosystems into agricultural fields and rangeland. Both of these processes reduce the amount of nitrogen stored in living vegetation and soil through the decomposition of organic matter.

Nitrous oxide is also released into the atmosphere when fossil fuels and biomass are burned. However, the combined contribution to the increase of this gas in the atmosphere is thought to be minor. The use of nitrate and ammonium fertilizers to enhance plant growth is another source of nitrous oxide. How much is released from this process has been difficult to quantify. Estimates suggest that the contribution from this source represents from 50% to 0.2% of nitrous oxide added to the atmosphere annually.

Ozone's role in the enhancement of the greenhouse effect has been difficult to determine. Accurate measurements of past long-term (more than 25 years in the past) levels of this gas in the atmosphere are currently unavailable. Moreover, concentrations of ozone gas are found in two different regions of the Earth's atmosphere. The majority of the ozone (about 97%) found in the atmosphere is concentrated in the stratosphere at an altitude of 15 to 55 kilometers above the Earth's surface.

This stratospheric ozone provides an important service to life on the Earth as it absorbs harmful ultraviolet radiation. In recent years, levels of stratospheric ozone have been decreasing due to the buildup of human created chlorofluorocarbons in the atmosphere.

Since the late 1970s, scientists have noticed the development of severe holes in the ozone layer over Antarctica. Satellite measurements have indicated that the zone from 65° North to 65° South latitude has had a 3% decrease in stratospheric ozone since 1978. Ozone is also highly concentrated at the Earth's surface in and around cities. Most of this ozone is created as a by product of human created photochemical smog. This buildup of ozone is toxic to organisms living at the Earth's surface.

THE LAYERED ATMOSPHERE

he Earth's atmosphere contains several different layers that can be defined according to air temperature, Figure 7b-1 displays these layers in an average atmosphere.

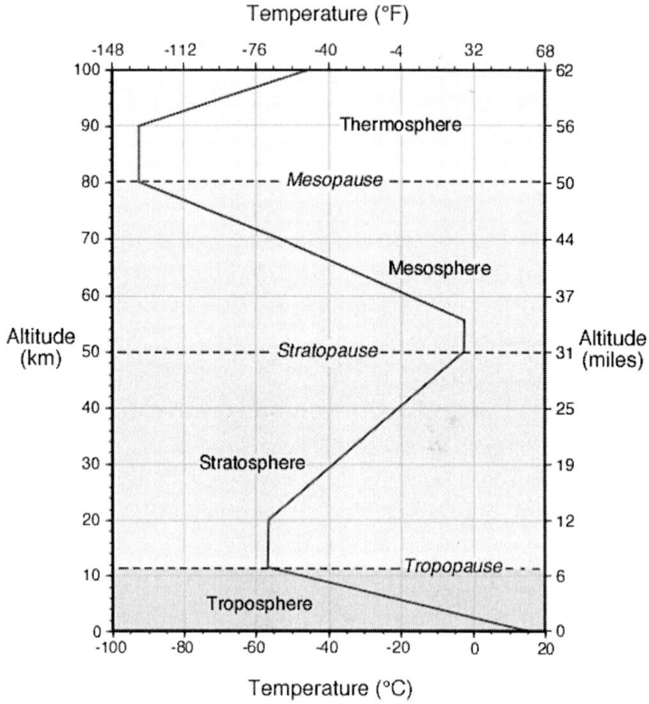

Fig. Vertical change in average global atmospheric temperature. Variations in the way temperature changes with height indicates the atmosphere is composed of a number of different layers. These variations are due to changes in the chemical and physical characteristics of the atmosphere with altitude.

According to temperature, the atmosphere contains four different layers. The first layer is called the troposphere. The depth of this layer varies from about 8 to 16 kilometers. Greatest depths occur at the tropics where warm temperatures causes vertical expansion of the lower atmosphere. From the tropics to the Earth's polar regions the troposphere becomes gradually thinner. The depth of this layer at the poles is roughly half as thick when compared to the tropics. Average depth of the troposphere is approximately 11 kilometers as displayed in Figure 7b-1.

About 80% of the total mass of the atmosphere is contained in troposphere. It is also the layer where the majority of our weather occurs. Maximum air temperature also occurs near the Earth's surface in this layer. With increasing height, air temperature drops uniformly with altitude at a rate of approximately 6.5° Celsius per 1000 meters.

This phenomenon is commonly called the Environmental Lapse Rate. At an average temperature of -56.5° Celsius, the top of the troposphere is reached. At the upper edge of the troposphere is a narrow transition zone known as the tropopause.

Above the tropopause is the stratosphere. This layer extends from an average altitude of 11 to 50 kilometers above the Earth's surface. This stratosphere contains about 19.9% of the total mass found in the atmosphere. Very little weather occurs in the stratosphere. Occasionally, the top portions of thunderstorms breach this layer. The lower portion of the stratosphere is also influenced by the polar jet stream and subtropical jet stream. In the first 9 kilometers of the stratosphere, temperature remains constant with height.

A zone with constant temperature in the atmosphere is called an isothermal layer. From an altitude of 20 to 50 kilometers, temperature increases with an increase in altitude.

The higher temperatures found in this region of the stratosphere occurs because of a localized concentration of ozone gas molecules. These molecules absorb ultraviolet sunlight creating heat energy that warms the stratosphere. Ozone is primarily found in the atmosphere at varying concentrations between the altitudes of 10 to 50 kilometers.

This layer of ozone is also called the ozone layer. The ozone layer is important to organisms at the Earth's surface as it protects them from the harmful effects of the Sun's ultraviolet radiation.

Without the ozone layer life could not exist on the Earth's surface.Separating the mesosphere from the stratosphere is transition zone called the stratopause. In the mesosphere, the atmosphere reaches its coldest temperatures (about -90° Celsius) at a height of approximately 80 kilometers. At the top of the mesosphere is another transition zone known as the mesopause.

The last atmospheric layer has an altitude greater than 80 kilometers and is called the thermosphere. Temperatures in this layer can be greater than 1200° C. These high temperatures are generated from the absorption of intense solar radiation by oxygen molecules (O2). While these temperatures seem extreme, the amount of heat energy involved is very small.

The amount of heat stored in a substance is controlled in part by its mass. The air in the thermosphere is extremely thin with individual gas molecules being separated from each other by large distances.

Consequently, measuring the temperature of thermosphere with a thermometer is a very difficult process. Thermometers measure the temperature of bodies via the movement of heat energy.

Normally, this process takes a few minutes for the conductive transfer of kinetic energy from countless molecules in the body of a substance to the expanding liquid inside the thermometer.

In the thermosphere, our thermometer would lose more heat energy from radiative emission then what it would gain from making occasional contact with extremely hot gas molecules.

PHYSICAL BEHAVIOUR OF THE ATMOSPHERE AND THE GAS LAWS

In the previous topic, we learned the atmosphere is composed of a mixture of many different gases. This mixture behaves in many ways as if it were a single gas. As a result of this phenomenon, the following generalizations describe important relationships between temperature, pressure, density and volume, that relate to the Earth's atmosphere.

- When temperature is held constant, the density of a gas is proportional to pressure, and volume is inversely proportional to pressure. Accordingly, an increase in pressure will cause an increase in density of the gas and a decrease in its volume.
- If volume is kept constant, the pressure of a unit mass of gas is proportional to temperature. If temperature increase so will pressure, assuming no change in the volume of the gas.
- Holding pressure constant, causes the temperature of a gas to be proportional to volume, and inversely proportional to density. Thus, increasing temperature of a unit mass of gas causes its volume to expand and its density to decrease as long as there is no change in pressure.

These relationships can also be described mathematically by the Ideal Gas Law. Two equations that are commonly used to describe this law are:

- Pressure x Volume = Constant x Temperature; and
- Pressure = Density x Constant x Temperature

ATMOSPHERIC PRESSURE

Air is a tangible material substance and as a result has mass. Any object with mass is influenced by the universal force known as gravity. Newton's Law of Universal Gravitation states: any two objects separated in space are attracted to each other by a force proportional to the product of their masses and inversely proportional to the square of the distance between them.

On the Earth, gravity can also be expressed as a force of acceleration of about 9.8 meters per second per second. As a result of this force, the speed of any object falling towards the surface of the Earth accelerates (1st second - 9.8 meters per second, 2nd second - 19.6 meters per second, 3rd second - 29.4 meters per second, and so on.) until terminal velocity is attained.

Gravity shapes and influences all atmospheric processes. It causes the density and pressure of air to decrease exponentially as one moves away from the surface of the Earth. Figure 7d-1 below models the average change in air pressure with height above the Earth's surface. In this graph, air pressure at the surface is illustrated as being approximately 1013 millibars (mb) or 1 kilogram per square centimeter of surface area.

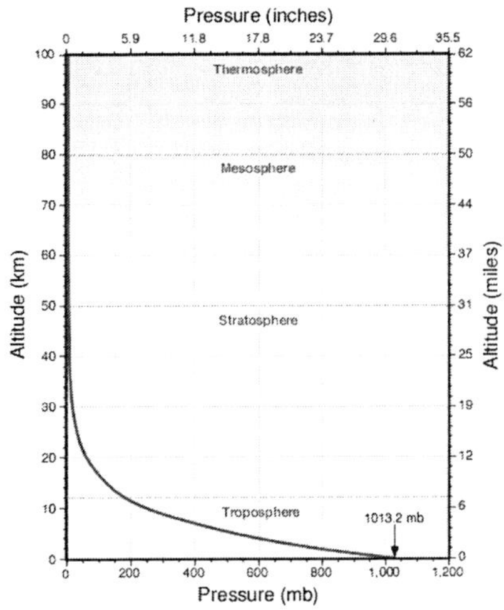

Fig. Change in average atmospheric pressure with altitude.

MEASURING ATMOSPHERIC PRESSURE

Any instrument that measures air pressure is called a barometer. The first measurement of atmospheric pressure began with a simple experiment performed by Evangelista Torricelli in 1643. In his experiment, Torricelli immersed a tube, sealed at one end, into a container of mercury. Atmospheric pressure then forced the mercury up into the tube to a level that was considerably higher than the mercury in the container. Torricelli determined from this experiment that the pressure of the atmosphere is approximately 30 inches or 76 centimeters (one centimeter of mercury is equal to 13.3 millibars). He also noticed that height of the mercury varied with changes in outside weather conditions.

Torricelli's Barometer

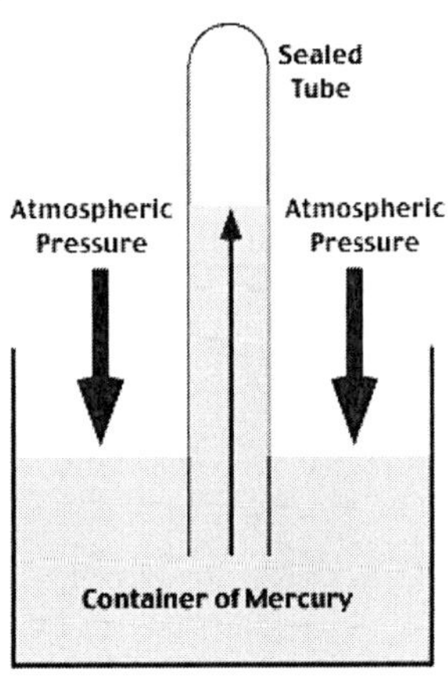

Fig. Diagram showing the construction of Torricelli's barometer.

The most common type barometer used in homes is the aneroid barometer. Inside this instrument is a small, flexible metal capsule called an aneroid cell. In the construction of the device, a vacuum is created inside the capsule so that small changes in outside air pressure cause the capsule to expand or contract.

The size of the aneroid cell is then calibrated and any change in its volume is transmitted by springs and levers to an indicating arm that points to the corresponding atmospheric pressure.

Fig. Aneroid barometer.

For climatological and meteorological purposes, standard sea-level pressure is said to be 76.0 cm or 29.92 inches or 1013.2 millibars. Scientists often use the kilopascal (kPa) as their preferred unit for measuring pressure. 1 kilopascal is equal to 10 millibars. Another unit of force sometimes used by scientists to measure atmospheric pressure is the newton. One millibar equals 100 newtons per square meter (N/m^2).

OZONE LAYER

The ozone layer is a layer in Earth's atmosphere which contains relatively high concentrations of ozone (O_3). This layer absorbs 97–99% of the Sun's high frequency ultraviolet light, which is damaging to life on Earth. It is mainly located in the lower portion of the stratosphere from approximately 13 to 40 kilometres (8.1 to 25 mi) above Earth, though the thickness varies seasonally and geographically. The ozone layer was discovered in 1913 by the French physicists Charles Fabry and Henri Buisson.

Its properties were explored in detail by the British meteorologist G. M. B. Dobson, who developed a simple spectrophotometer (the Dobsonmeter) that could be used to measure stratospheric ozone from the ground. Between 1928 and 1958 Dobson established a worldwide network of ozone monitoring stations, which continue to operate to this day. The "Dobson unit", a convenient measure of the columnar density of ozone overhead, is named in his Honour.

ORIGIN OF OZONE

The photochemical mechanisms that give rise to the ozone layer were discovered by the British physicist Sidney Chapman in 1930. Ozone in the Earth's stratosphere is created by ultraviolet light striking oxygen molecules containing two oxygen atoms (O_2), splitting them into individual oxygen atoms (atomic oxygen); the atomic oxygen then combines with unbroken O_2 to create ozone, O_3. The ozone molecule is also unstable (although, in the stratosphere, long-lived) and when ultraviolet light hits ozone it splits into a molecule of O_2 and an atom of atomic oxygen, a continuing process called the ozone-oxygen cycle, thus creating an ozone layer in the stratosphere, the region from about 10 to 50 kilometres (33,000 to 160,000 ft) above Earth's surface. About 90% of the ozone in our atmosphere is contained in the stratosphere. Ozone concentrations are greatest between about 20 and 40 kilometres (12 and 25 mi), where they range from about 2 to 8 parts per million. If all of the ozone were compressed to the pressure of the air at sea level, it would be only a few millimeters thick.

ULTRAVIOLET LIGHT AND OZONE

Although the concentration of the ozone in the ozone layer is very small, it is vitally important to life because it absorbs biologically harmful ultraviolet (UV) radiation coming from the sun. UV radiation is divided into three categories, based on its wavelength; these are referred to as UV-A (400–315 nm), UV-B (315–280 nm), and UV-C (280–100 nm). UV-C, which would be very harmful to all living things, is entirely screened out by ozone at around 35 kilometres (115,000 ft) altitude. UV-B radiation can be harmful to the skin and is the main cause of sunburn; excessive exposure can also cause genetic damage, resulting in problems such as skin cancer. The ozone layer is very effective at screening out UV-B; for radiation with a wavelength of 290 nm, the intensity at the top of the atmosphere is 350 million times stronger than at the Earth's surface. Nevertheless, some UV-B reaches the surface. Most UV-A reaches the surface; this radiation is significantly less harmful, although it can potentially cause genetic damages.

DISTRIBUTION OF OZONE IN THE STRATOSPHERE

The thickness of the ozone layer—that is, the total amount of ozone in a column overhead—varies by a large factor worldwide, being in general smaller near the equator and larger towards the poles. It also varies with season, being in general thicker during the spring and thinner during the autumn in the northern hemisphere. The reasons for this latitude and seasonal dependence are complicated, involving atmospheric circulation patterns as well as solar intensity.Since stratospheric ozone is produced by solar UV radiation, one might expect to find the highest ozone levels over the tropics and the lowest over polar regions. The same argument would lead one to expect the highest ozone levels in the summer and the lowest in the winter. The observed Behaviour is very different: most of the ozone is found in the mid-to-high latitudes of the northern and southern hemispheres, and the highest levels are found in the spring, not summer, and the lowest in the autumn, not winter in the northern hemisphere. During winter, the ozone layer actually increases in depth. This puzzle is explained by the prevailing stratospheric wind patterns, known as the Brewer-Dobson circulation. While most of the ozone is indeed created over the tropics, the stratospheric circulation then transports it poleward and downward to the lower stratosphere of the high latitudes. However in the southern hemisphere, owing to the ozone hole phenomenon, the lowest amounts of column ozone found anywhere in the world are over the Antarctic in the southern spring period of September and October.

The ozone layer is higher in altitude in the tropics, and lower in altitude in the extratropics, especially in the polar regions. This altitude variation of ozone results from the slow circulation that lifts the ozone-poor air out of the troposphere into the stratosphere. As this air slowly rises in the tropics, ozone is produced by the overhead sun which photolyzes oxygen molecules. As this slow circulation bends towards the mid-latitudes, it carries the ozone-rich air from the tropical middle stratosphere to the mid-and-high latitudes lower stratosphere. The high ozone concentrations at high latitudes are due to the accumulation of ozone at lower altitudes.The Brewer-Dobson circulation moves very slowly. The time needed to lift an air parcel from the tropical tropopause near 16 to 20 kilometres (9.9 to 12 mi) is about 4–5 months (about 30 feet (9.1 m) per day). Even though ozone in the lower tropical stratosphere is produced at a very slow rate, the lifting circulation is so slow that ozone can build up to relatively high levels by the time it reaches 26 kilometres (16 mi).Ozone amounts over the continental United States (25°N to 49°N) are highest in the northern spring (April and May). These ozone amounts fall over the course of the summer to their lowest amounts in October, and then rise again over the course of the winter. Again, wind transport of ozone is principally responsible for the seasonal evolution of these higher latitude ozone patterns.

The total column amount of ozone generally increases as we move from the tropics to higher latitudes in both hemispheres. However, the overall column amounts are greater in the northern hemisphere high latitudes than in the southern hemisphere high latitudes. In addition, while the highest amounts of column ozone over the Arctic occur in the northern spring (March–April), the opposite is true over the Antarctic, where the lowest amounts of column ozone occur in the southern spring (September–October).

OZONE DEPLETION

The ozone layer can be depleted by free radical catalysts, including nitric oxide (NO), nitrous oxide (N_2O), hydroxyl (OH), atomic chlorine (Cl), and atomic bromine (Br). While there are natural sources for all of these species, the concentrations of chlorine and bromine have increased markedly in recent years due to the release of large quantities of man-made organohalogen compounds, especially chlorofluorocarbons (CFCs) and bromofluorocarbons. These highly stable compounds are capable of surviving the rise to the stratosphere, where Cl and Br radicals are liberated by the action of ultraviolet light. Each radical is then free to initiate and catalyze a chain reaction capable of breaking down over 100,000 ozone molecules. The breakdown of ozone in the stratosphere results in the ozone molecules being unable to absorb ultraviolet radiation. Consequently, unabsorbed and dangerous ultraviolet-B radiation is able to reach the Earth's surface. Ozone levels, over the northern hemisphere, have been dropping by 4% per decade. Over approximately 5% of the Earth's surface, around the north and south poles, much larger (but seasonal) declines have been seen; these are the ozone holes. In 2009, nitrous oxide (N_2O) was the largest ozone-depleting substance emitted through human activities.

Regulation

In 1978, the United States, Canada and Norway enacted bans on CFC-containing aerosol sprays that are thought to damage the ozone layer. The European Community rejected an analogous proposal to do the same. In the U.S., chlorofluorocarbons continued to be used in other applications, such as refrigeration and industrial cleaning, until after the discovery of the Antarctic ozone hole in 1985. After negotiation of an international treaty (the Montreal Protocol), CFC production was sharply limited beginning in 1987 and phased out completely by 1996. Since that time, the treaty has been amended to ban CFC production after 1995 in the developed countries, and later in developing. Today, over 160 countries have signed the treaty. Beginning January 1, 1996, only recycled and stockpiled CFCs will be available for use in developed countries like the US. This production phaseout is possible because of efforts to ensure that there will be substitute chemicals and technologies for all CFC uses. On August 2, 2003, scientists announced that the depletion of the ozone layer may be slowing down due to the international ban on CFCs. Three satellites and three ground stations confirmed that the upper atmosphere ozone depletion rate has slowed down significantly during the past decade. The study was organized by the American Geophysical Union. Some breakdown can be expected to continue due to CFCs used by nations which have not banned them, and due to gases which are already in the stratosphere. CFCs have very long atmospheric lifetimes, ranging from 50 to over 100 years, so the final recovery of the ozone layer is expected to require several lifetimes.Compounds containing C–H bonds (such as hydrochlorofluorocarbons, or HCFCs) have been designed to replace the function of CFCs. These replacement compounds are more reactive and less likely to survive long enough in the atmosphere to reach the stratosphere where they could affect the ozone layer. While being less damaging than CFCs, HCFCs can have a negative impact on the ozone layer, so they are also being phased out.

ATMOSPHERIC EFFECTS ON INCOMING SOLAR RADIATION

Three atmospheric processes modify the solar radiation passing through our atmosphere destined to the Earth's surface. These processes act on the radiation when it interacts with gases and suspended particles found in the atmosphere.

The process of scattering occurs when small particles and gas molecules diffuse part of the incoming solar radiation in random directions without any alteration to the wavelength of the electromagnetic energy. Scattering does, however, reduce the amount of incoming radiation reaching the Earth's surface. A significant proportion of scattered shortwave solar radiation is redirected back to space. The amount of scattering that takes place is dependent on two factors: wavelength of the incoming radiation and the size of the scattering particle or gas molecule. In the Earth's atmosphere, the presence of a large number of particles with a size of about 0.5 microns results in shorter wavelengths being preferentially scattered. This factor also causes our sky to look blue because this Colour corresponds to those wavelengths that are best diffused. If scattering did not occur in our atmosphere the daylight sky would be black.

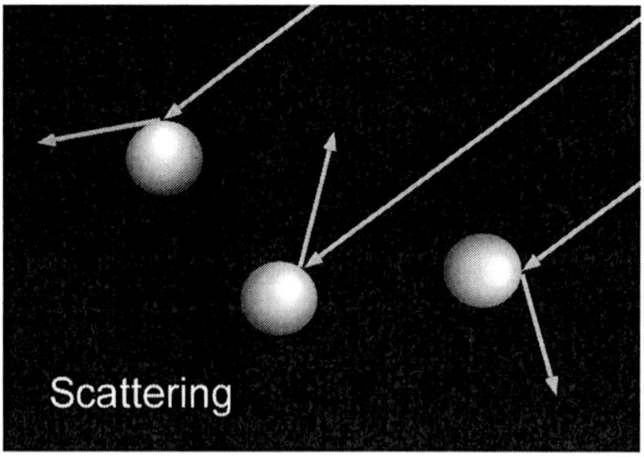

Fig. The process of atmospheric scattering causes rays of sunlight to be redirected to a new direction after hitting a particle in the atmosphere. In this illustration, we see how three particles send light rays off into three different directions. Scattering does not change the striking light ray's wavelength or intensity.

If intercepted, some gases and particles in the atmosphere have the ability to absorb incoming insolation. Absorption is defined as a process in which solar radiation is retained by a substance and converted into heat energy. The creation of heat energy also causes the substance to emit its own radiation. In general, the absorption of solar radiation by substances in the Earth's atmosphere results in temperatures that get no higher than 1800° Celsius. According to Wien's Law, bodies with temperatures at this level or lower would emit their radiation in the longwave band. Further, this emission of radiation is in all directions so a sizable proportion of this energy is lost to space.

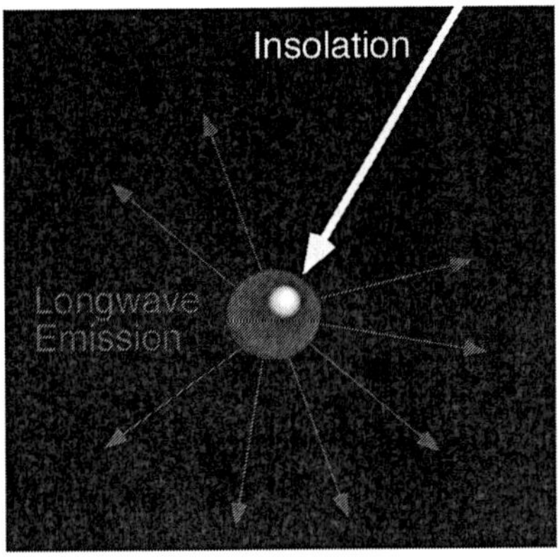

Fig. Atmospheric absorption. In this process, sunlight is absorbed by an atmospheric particle, transferred into heat energy, and then converted into longwave radiation emissions that come from the particle.

The final process in the atmosphere that modifies incoming solar radiation is reflection. Reflection is a process where sunlight is redirect by 180° after it strikes an atmospheric particle.

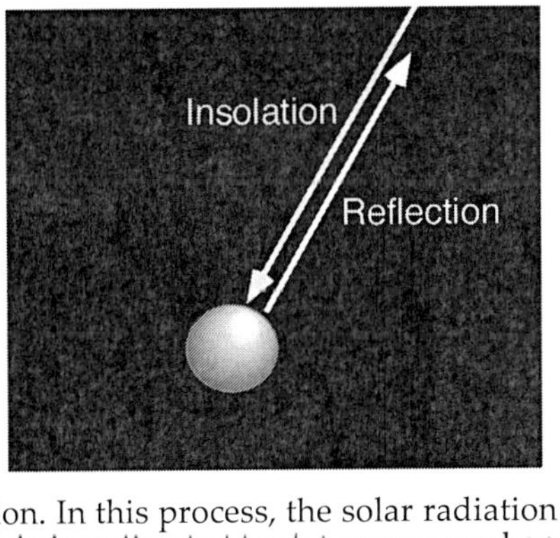

Fig. Atmospheric reflection. In this process, the solar radiation striking an atmospheric particle is redirected back to space unchanged.

This redirection causes a 100% loss of the insolation. Most of the reflection in our atmosphere occurs in clouds when light is intercepted by particles of liquid and frozen water. The reflectivity of a cloud can range from 40 to 90%.

Sunlight reaching the Earth's surface unmodified by any of the above atmospheric processes is termed direct solar radiation. Solar radiation that reaches the Earth's surface after it was altered by the process of scattering is called diffused solar radiation. Not all of the direct and diffused radiation available at the Earth's surface is used to do *work* (photosynthesis, creation of sensible heat, evaporation, etc.). As in the atmosphere, some of the radiation received at the Earth's surface is redirected back to space by reflection. The following image describes the spatial pattern of surface reflectivity as measured for the year 1987.

The reflectivity or albedo of the Earth's surface varies with the type of material that covers it. For example, fresh snow can reflect up to 95% of the insolation that reaches it surface. Some other surface type reflectivities are:

- Dry sand 35 to 45%
- Broadleaf deciduous forest 5 to 10%
- Needleleaf confierous forest 10 to 20%
- Grass type vegetation 15 to 25%

Reflectivity of the surface is often described by the term *surface* albedo. The Earth's average *albedo*, reflectance from both the atmosphere and the surface, is about 30%.

Figure 7f-5 describes the modification of solar radiation by atmospheric and surface processes for the whole Earth over a period of one year. Of all the sunlight that passes through the atmosphere annually, only 51% is available at the Earth's surface to do work. This energy is used to heat the Earth's surface and lower atmosphere, melt and evaporate water, and run photosynthesis in plants. Of the other 49%, 4% is reflected back to space by the Earth's surface, 26% is scattered or reflected to space by clouds and atmospheric particles, and 19% is absorbed by atmospheric gases, particles, and clouds.

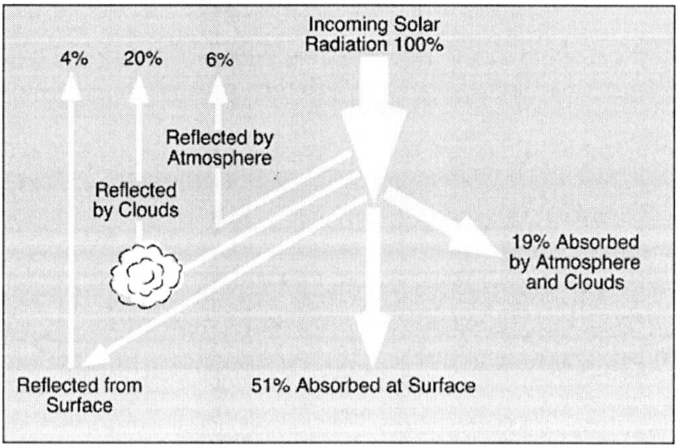

Fig. Global modification of incoming solar radiation by atmospheric and surface processes.

GLOBAL PATTERNS OF INSOLATION RECEIPTS

The combined effect of Earth-Sun relationships (angle of incidence and day length variations) and the modification of the solar beam as it passes through the atmosphere produces specific global patterns of annual insolation receipt as seen on Figure 7g-1 above (and see the *NASA WWW links* below). After examining these patterns, the following trends can be identified:

- Highest values of insolation received occur in tropical latitudes. Within this zone there are localized maximums over the tropical oceans and deserts where the atmosphere has virtually no cloud development for most of the year. Insolation quantities at the equator over land during the solstices are approximately the same as values found in the middle latitudes during their summer.
- Outside the tropics, annual receipts of solar radiation generally decrease with increasing latitude. Minimum values occur at the poles. This pattern is primarily the result of Earth-Sun geometric relationships and its effect on the duration and intensity of solar radiation received.
- In middle and high latitudes, insolation values over the ocean, as compared to those at the same latitude over the land, are generally higher. Greater cloudiness over land surfaces accounts for this variation.

In the equations above, the mathematical terms have the following definitions:

- K = Shortwave Direct Radiation
- k = Shortwave Indirect Radiation
- a = Reflectivity of the Surface or Surface Albedo

GREENHOUSE EFFECT

The greenhouse effect is a process by which thermal radiation from a planetary surface is absorbed by atmospheric greenhouse gases, and is re-radiated in all directions.

Since part of this re-radiation is back towards the surface, energy is transferred to the surface and the lower atmosphere. As a result, the temperature there is higher than it would be if direct heating by solar radiation were the only warming mechanism.

This mechanism is fundamentally different from that of an actual greenhouse, which works by isolating warm air inside the structure so that heat is not lost by convection.

The greenhouse effect was discovered by Joseph Fourier in 1824, first reliably experimented on by John Tyndall in 1858, and first reported quantitatively by Svante Arrhenius in 1896. If an ideal thermally conductive blackbody was the same distance from the Sun as the Earth is, it would have a temperature of about 5.3 °C. However, since the Earth reflects about 30% (or 28%) of the incoming sunlight, the planet's effective temperature (the temperature of a blackbody that would emit the same amount of radiation) is about "18 or "19 °C, about 33°C below the actual surface temperature of about 14 °C or 15 °C. The mechanism that produces this difference between the actual surface temperature and the effective temperature is due to the atmosphere and is known as the greenhouse effect.

Global warming, a recent warming of the Earth's surface and lower atmosphere, is believed to be the result of a strengthening of the greenhouse effect mostly due to human-produced increases in atmospheric greenhouse gases.

BASIC MECHANISM

The Earth receives energy from the Sun in the form UV, visible, and near IR radiation, most of which passes through the atmosphere without being absorbed. Of the total amount of energy available at the top of the atmosphere (TOA), about 50% is absorbed at the Earth's surface. Because it is warm, the surface radiates far IR thermal radiation that consists of wavelengths that are predominantly much longer than the wavelengths that were absorbed. Most of this thermal radiation is absorbed by the atmosphere and re-radiated both upwards and downwards; that radiated downwards is absorbed by the Earth's surface. This trapping of long-wavelength thermal radiation leads to a higher equilibrium temperature than if the atmosphere were absent.

This highly simplified picture of the basic mechanism needs to be qualified in a number of ways, none of which affect the fundamental process.

- The incoming radiation from the Sun is mostly in the form of visible light and nearby wavelengths, largely in the range 0.2- 4 ìm, corresponding to the Sun's radiative temperature of 6,000 K. Almost half the radiation is in the form of "visible" light, which our eyes are adapted to use.
- About 50% of the Sun's energy is absorbed at the Earth's surface and the rest is reflected or absorbed by the atmosphere. The reflection of light back into space—largely by clouds—does not much affect the basic mechanism; this light, effectively, is lost to the system.
- The absorbed energy warms the surface. Simple presentations of the greenhouse effect, such as the idealized greenhouse model, show this heat being lost as thermal radiation. The reality is more complex: the atmosphere near the surface is largely opaque to thermal radiation (with important exceptions for "window" bands), and most heat loss from the surface is by sensible heat and latent heat transport. Radiative energy losses become increasingly important higher in the atmosphere largely because of the decreasing concentration of water vapor, an

important greenhouse gas. It is more realistic to think of the greenhouse effect as applying to a "surface" in the mid-troposphere, which is effectively coupled to the surface by a lapse rate.

- Within the region where radiative effects are important the description given by the idealized greenhouse model becomes realistic: The surface of the Earth, warmed to a temperature around 255 K, radiates long-wavelength, infrared heat in the range 4–100 μm. At these wavelengths, greenhouse gases that were largely transparent to incoming solar radiation are more absorbent. Each layer of atmosphere with greenhouses gases absorbs some of the heat being radiated upwards from lower layers. To maintain its own equilibrium, it re-radiates the absorbed heat in all directions, both upwards and downwards. This results in more warmth below, while still radiating enough heat back out into deep space from the upper layers to maintain overall thermal equilibrium. Increasing the concentration of the gases increases the amount of absorption and re-radiation, and thereby further warms the layers and ultimately the surface below.
- Greenhouse gases—including most diatomic gases with two different atoms (such as carbon monoxide, CO) and all gases with three or more atoms—are able to absorb and emit infrared radiation. Though more than 99% of the dry atmosphere is IR transparent (because the main constituents—N_2, O_2, and Ar—are not able to directly absorb or emit infrared radiation), intermolecular collisions cause the energy absorbed and emitted by the greenhouse gases to be shared with the other, non-IR-active, gases.
- The simple picture assumes equilibrium. In the real world there is the diurnal cycle as well as seasonal cycles and weather. Solar heating only applies during daytime. During the night, the atmosphere cools somewhat, but not greatly, because its emissivity is low, and during the day the atmosphere warms. Diurnal temperature changes decrease with height in the atmosphere.

GREENHOUSE GASES

By their percentage contribution to the greenhouse effect on Earth the four major gases are:

- Water vapor, 36–70%
- Carbon dioxide, 9–26%
- Methane, 4–9%
- Ozone, 3–7%

The major non-gas contributor to the Earth's greenhouse effect, clouds, also absorb and emit infrared radiation and thus have an effect on radiative properties of the atmosphere.

ROLE IN CLIMATE CHANGE

Strengthening of the greenhouse effect through human activities is known as the enhanced (or anthropogenic) greenhouse effect. This increase in radiative forcing from human activity is attributable mainly to increased atmospheric carbon dioxide levels.CO_2 is produced by fossil fuel burning and other activities such as cement production and tropical deforestation. Measurements of CO_2 from the Mauna Loa observatory show that concentrations have increased from about 313 ppm in 1960 to about 389 ppm in 2010. The current observed amount of CO_2 exceeds the geological record maxima (~300 ppm) from ice core data. The effect of combustion-produced carbon dioxide on the global climate, a special case of the greenhouse effect first described in 1896 by Svante Arrhenius, has also been called the Callendar effect.

Because it is a greenhouse gas, elevated CO_2 levels contribute to additional absorption and emission of thermal infrared in the atmosphere, which produce net warming. According to the latest Assessment Report from the Intergovernmental Panel on Climate Change, "most of the observed increase in globally averaged temperatures since the mid-20th century is very likely due to the observed increase in anthropogenic greenhouse gas concentrations".

Over the past 800,000 years, ice core data shows unambiguously that carbon dioxide has varied from values as low as 180 parts per million (ppm) to the pre-industrial level of 270ppm. Paleoclimatologists consider variations in carbon dioxide to be a fundamental factor in controlling climate variations over this time scale.

REAL GREENHOUSES

The "greenhouse effect" is named by analogy to greenhouses. The greenhouse effect and a real greenhouse are similar in that they both limit the rate of thermal energy flowing out of the system, but the mechanisms by which heat is retained are different. A greenhouse works primarily by preventing absorbed heat from leaving the structure through convection, *i.e.* sensible heat transport. The greenhouse effect heats the earth because greenhouse gases absorb outgoing radiative energy and re-emit some of it back towards earth.

A greenhouse is built of any material that passes sunlight, usually glass, or plastic. It mainly heats up because the Sun warms the ground inside, which then warms the air in the greenhouse. The air continues to heat because it is confined within the greenhouse, unlike the environment outside the greenhouse where warm air near the surface rises and mixes with cooler air aloft. This can be demonstrated by opening a small window near the roof of a greenhouse: the temperature will drop considerably. It has also been demonstrated experimentally that a "greenhouse" with a cover of rock salt heats up an enclosure similarly to one with a glass cover. Thus greenhouses work primarily by preventing convective cooling.

In the greenhouse effect, rather than retaining (sensible) heat by physically preventing movement of the air, greenhouse gases act to warm the Earth by re-radiating some of the energy back towards the surface. This process may exist in real greenhouses, but is comparatively unimportant there.

BODIES OTHER THAN EARTH

In our solar system, Mars, Venus, and the moon Titan also exhibit greenhouse effects. Titan has an anti-greenhouse effect, in that its atmosphere absorbs solar radiation but is relatively transparent to infrared radiation. Pluto also exhibits Behaviour superficially similar to the anti-greenhouse effect. A runaway greenhouse effect occurs if positive feedbacks lead to the evaporation of all greenhouse gases into the atmosphere. A runaway greenhouse effect involving carbon dioxide and water vapor is thought to have occurred on Venus.

NET RADIATION AND THE PLANETARY ENERGY BALANCE

Shortwave radiation from the Sun enters the surface-atmosphere system of the Earth and is ultimately returned to space as longwave radiation (because the Earth is cooler than the Sun). A basic necessity of this energy interchange is that incoming solar insolation and outgoing radiation be equal in quantity. One way of modeling this balance in energy exchange is described graphically with the use of the following two *cascade* diagrams.

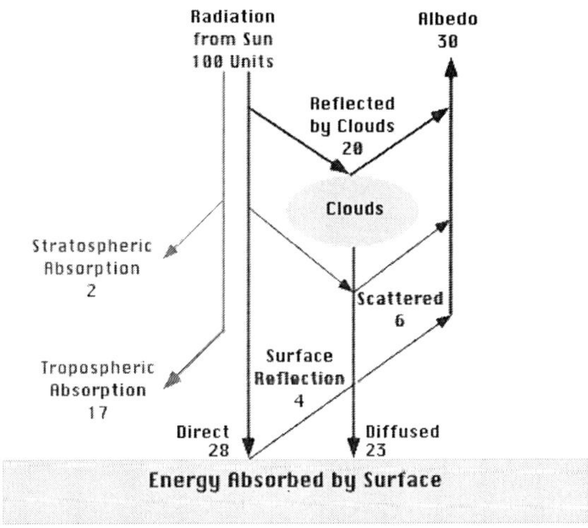

Fig. Global shortwave radiation cascade.

The *Global Shortwave Radiation Cascade* describes the relative amounts (based on 100 units available at the top of the atmosphere) of shortwave radiation partitioned to various atmospheric processes as it passes through the atmosphere. The diagram indicates that 19 units of insolation are absorbed (and therefore transferred into heat energy and longwave radiation) in the atmosphere by the following two processes:

- Stratospheric Absorption of the Ultraviolet Radiation by Ozone 2 units; and
- Tropospheric Absorption of Insolation by Clouds and Aerosols 17 units. 23 units of solar radiation are scattered in the atmosphere subsequently absorbed at the surface as diffused insolation. 28 units of the incoming solar radiation are absorbed at the surface as direct insolation. Total amount of solar insolation absorbed at the surface equals 51 units. The total amount of shortwave radiation absorbed at the surface and in the atmosphere is 70 units.

Three main losses of solar radiation back to space occur in the Earth's shortwave radiation cascade. 4 units of sunlight are returned to space from surface reflection. Cloud reflection returns another 20 units of solar radiation. Back scattering of sunlight returns 6 units to space. The total loss of shortwave radiation from these processes is 30 units. The term used to describe the combined effect of all of these shortwave losses is Earth albedo.

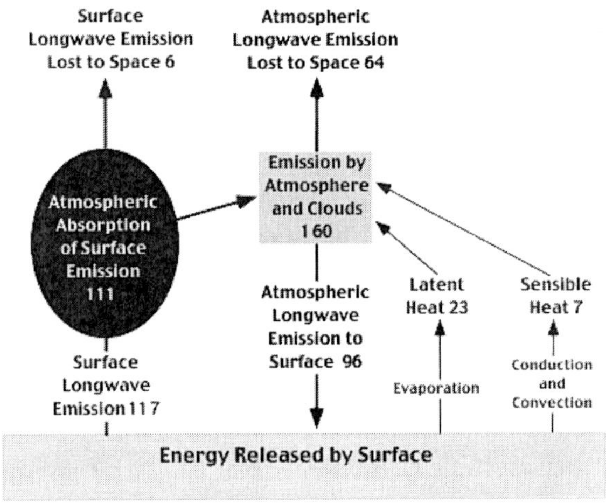

Fig. Global longwave radiation cascade.

The *Global Longwave Radiation Cascade* indicates that energy leaves the Earth's surface through three different processes. 7 units leave the surface as sensible heat. This heat is transferred into the atmosphere by conduction and convection. The melting and evaporation of water at the Earth's surface incorporates 23 units energy into the atmosphere as latent heat. This latent heat is released into the atmosphere when the water condenses or becomes solid. Both of these processes become part of the emission of longwave radiation by the atmosphere and clouds.

The surface of the Earth emits 117 units of longwave radiation. Of this emission only 6 units are directly lost to space. The other 111 units are absorbed by greenhouse gases in the atmosphere and converted into heat energy and then into atmospheric emissions of longwave radiation (the greenhouse effect).

The atmosphere emits 160 units of longwave energy. Contributions to this 160 units are from surface emissions of longwave radiation (111 units), latent heat transfer (23 units), sensible heat transfer (7 units), and the absorption of shortwave radiation by atmospheric gases and clouds. Atmospheric emissions travel in two directions. 64 units of atmospheric emission is lost directly to space. 96 units travel to the Earth's surface where it is absorbed and transferred into heat energy.

The total amount of energy lost to space in the global longwave radiation cascade is 70 units (surface emission 6 units + atmospheric emission 64 units.)

This is the same amount of energy that was added to the Earth's atmosphere and surface by the *Global Shortwave Radiation Cascade*.

Finally, to balance the surface energy exchanges in this cascade we have to account for 51 units of missing energy [atmosphere and cloud longwave emission (96 units) minus surface longwave emission (117 units) minus latent heat transfer (23 units) minus sensible heat transfer (7 units) = -51 units]. This missing component to the radiation balance is the 51 units of energy absorbed at the Earth's surface as direct and diffused shortwave radiation.

The following equations can be used to mathematically model net shortwave radiation balance, net longwave radiation balance, and net radiation balance for the Earth's surface at a single location or for the whole globe for any temporal period:

- $K^* = (K + k)(1 - a)$
- $L^* = (LD - LU)$
- $Q^* = (K + k)(1 - a) - LU + LD$

where

- Q^* is surface net radiation (global annual values of $Q^* = 0$, because input equals output, local values can be positive or negative),
- K^* is surface net shortwave radiation,
- K is surface direct shortwave radiation,
- k is diffused shortwave radiation (scattered insolation) at the surface,
- a is the albedo of surface,
- L^* is net longwave radiation at the surface,
- LD is atmospheric counter-radiation directed to the Earth's surface, and
- LU is longwave radiation lost from the Earth's surface.

GLOBAL HEAT BALANCE: INTRODUCTION TO HEAT FLUXES

Figure illustrates the annual values of net shortwave and net longwave radiation from the South Pole to the North Pole. On closer examination of this graph one notes that the lines representing *incoming* and *outgoing radiation* do not have the same values. From 0 - 35 ° latitude North and South incoming solar radiation exceeds outgoing terrestrial radiation and a surplus of energy exists. The reverse holds true from 35 - 90° latitude North and South and these regions have a deficit of energy. Surplus energy at low latitudes and a deficit at high latitudes results in energy transfer from the equator to the poles. It is this meridional transport of energy that causes atmospheric and oceanic circulation. If there were no energy transfer the poles would be 25° Celsius cooler, and the equator 14° Celsius warmer!

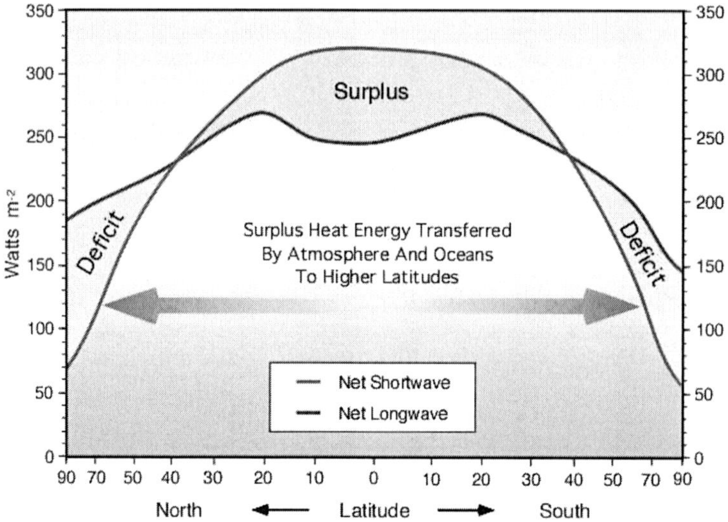

Fig. Balance between average net shortwave and longwave radiation from 90° North to 90° South.

The redistribution of energy across the Earth's surface is accomplished primarily through three processes: sensible heat flux, latent heat flux, and surface heat flux into oceans. Sensible heat flux is the process where heat energy is transferred from the Earth's surface to the atmosphere by conduction and convection.

This energy is then moved from the tropics to the poles by advection, creating atmospheric circulation. As a result, atmospheric circulation moves warm tropical air to the polar regions and cold air from the poles to the equator. Latent heat flux moves energy globally when solid and liquid water is converted into vapor.

This vapor is often moved by atmospheric circulation vertically and horizontally to cooler locations where it is condensed as rain or is deposited as snow releasing the heat energy stored within it. Finally, large quantities of radiation energy are transferred into the Earth's tropical oceans. The energy enters these water bodies at the surface when absorbed radiation is converted into heat energy. The warmed surface water is then transferred downward into the water column by conduction and convection. Horizontal transfer of this heat energy from the equator to the poles is accomplished by ocean currents.

The following equation describes the partitioning of heat energy at the Earth's surface:

- Q* = H (Sensible heat) + L (Latent heat) + S (Surface heat flux into soil or water)

The actual amount of net radiation being partitioned into each one of these components is a function of the following factors:

- Presence or absence of water in liquid and solid forms at the surface.
- Specific heat of the surface receiving the net radiation.
- Convective and conductive characteristics of the receiving surface.
- Diffusion characteristics of the surface's overlying atmosphere.

THE CONCEPT OF TEMPERATURE

TEMPERATURE AND HEAT

Temperature and heat are not the same phenomenon. Temperature is a measure of the intensity or degree of hotness in a body. Technically, it is determined by getting the average speed of a body's molecules. Heat is a measure of the quantity of heat energy present in a body. The spatial distribution of temperature in a body determines heat flow. Heat always flows from warmer to colder areas.

The heat held in a object depends not only on its temperature but also its mass. For example, let us compare the heating of two different masses of water. In this example, one mass has a weight of 5 grams, while the other is 25 grams. If the temperature of both masses is raised from 20 to 25° Celsius, the larger mass of water will require five times more heat energy for this increase in temperature.

This larger mass would also contain contain 5 times more stored heat energy.

Table. Heat energy required to raise two different quantities of water 5° Celsius.

Mass of the Water	Starting Temperature	Ending Temperature	Heat Required
5 grams	20° Celsius	25° Celsius	25 Calories of Heat
25 grams	20° Celsius	25° Celsius Heat	125 Calories of

TEMPERATURE SCALES

A number of measurement scales have been invented to measure temperature. Table 7k-2 describes important temperatures for the three dominant scales in use today.

Table. Temperature of absolute zero, the ice point of water, and the stream point of water using various temperature measurement scales.

Measurement Scale	Steam Point of Water	Ice Point of Water	Absolute Zero
Fahrenheit	212	32	-460
Celsius	100	0	-273
Kelvin	373	273	0

The most commonly used scale for measuring temperature is the Celsius system. The Celsius scale was developed in 1742 by the Swedish astronomer Anders Celsius. In this system, the melting point of ice was given a value of 0, the boiling point of water is 100, and absolute zero is -273.

The Fahrenheit system is a temperature scale that is used exclusively in the United States. This system was created by German physicist Gabriel Fahrenheit in 1714. In this scale, the melting point of ice has a value of 32, water boils at 212, and absolute zero has a temperature of -460. The Kelvin scale was proposed by British physicist Lord Kelvin in 1848.

This system is often used by scientists because its temperature readings begin at absolute zero and due to the fact that this scale is proportional to the amount of heat energy found in an object. The Kelvin scale assigns a value of 273 for the melting temperature of ice, while the boiling point of water occurs at 373.

MEASUREMENT OF AIR TEMPERATURE

A thermometer is a device that is used to measure temperature. Thermometers consist of a sealed hollow glass tube filled with some type of liquid. Thermometers measure temperature by the change in the volume of the liquid as it responds to the addition or loss of heat energy from the environment immediately outside its surface.

When heat is added, the liquid inside the thermometer expands. Cooling cause the liquid to contract. Meteorological thermometers are often filled with either alcohol or mercury. Alcohol thermometers are favored in very cold environments because of this liquid's low freezing point (-112° Celsius).

By international agreement, the nations of the world have decided to measure temperature in a similar fashion. This standardization is important for the accurate generation of weather maps and forecasts, both of which depend on having data determined in a uniform way.

Weather stations worldwide try to determine minimum and maximum temperatures for each and every day. By averaging these two values, daily mean temperatures are also calculated. Many stations also take temperature readings on the hour. Temperature measurements are determined by thermometers designed and approved by the World Meteorological Organization. These instruments are housed in specially designed instrument shelters that allow for the standardization of measurements taken anywhere on the Earth.

Fig. Well ventilated instrument shelters are used to protect thermometers from precipitation, direct Sun, and other physical elements. Construction standardization of these shelters, by international agreement, guarantees that measurements are comparable in any of the over 15,000 weather stations found worldwide.

Fig. Thermometers found inside the instrument shelter are mounted approximate 1.5 meters above the ground surface. The top thermometer contains alcohol and is used to determine daily minimum temperatures. The lower thermometer uses mercury to determine the daily maximum temperature.

DAILY AND ANNUAL CYCLES OF TEMPERATURE

DAILY CYCLES OF AIR TEMPERATURE

At the Earth's surface quantities of insolation and net radiation undergo daily cycles of change because the planet rotates on its polar axis once every 24 hours. Insolation is usually the main positive component making up net radiation. Variations in net radiation are primarily responsible for the particular patterns of rising and falling air temperature over a 24 hour period. The following three graphs show hypothetical average curves of *insolation*, *net radiation*, and *air temperature* for a typical land based location at 45° of latitude on the equinoxes and solstices.

Insolation

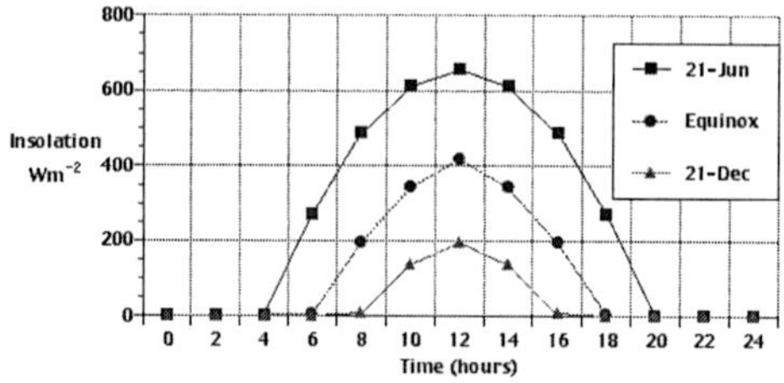

Fig. Hourly variations in insolation received for a location at 45° North latitude over a 24 hour period.

In the above graph, shortwave radiation received from the Sun is measured in Watts. For all dates, peak reception occurs at solar noon when the Sun attains its greatest height above the horizon.

Net Radiation

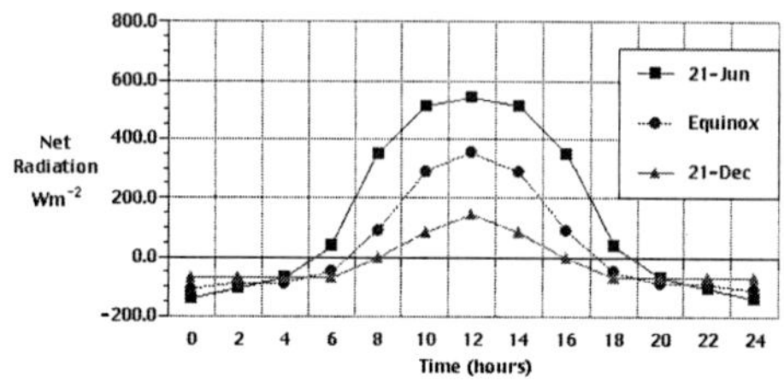

Fig. Hourly variations in net radiation for a location at 45° North latitude over a 24 hour period.

Units in Figure 71-2 are the same as the insolation graph above. The net radiation graph indicates that there is a surplus of radiation during most of the day and a deficit throughout the night. The deficit begins just before sunset when emitted longwave radiation from the Earth's surface exceeds solar insolation and longwave radiation from the atmosphere.

Temperature

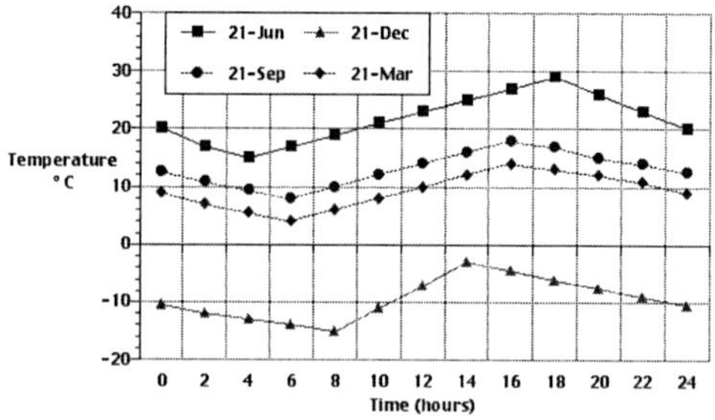

Fig. Hourly variations in surface temperature for a location at 45° North latitude over a 24 hour period.

The relative placement of the temperature profiles for the various dates correlates to the amount of net radiation available for daily surface absorption and heat generation. The more energy available, the higher up the Y-axis the profile is on the graph. September equinox is warmer than the March equinox because of the heating that occurred in the previous summer months. For all dates, minimum temperature occurs at *sunrise*. Temperature drops throughout the night because of two processes. First, the Earth's radiation balance at the surface becomes negative after *sunset*. Thus, the surface of the Earth stops heating up as solar radiation is not being absorbed. Secondly, conduction and convection transport heat energy up into the atmosphere and the warm air that was at the surface is replaced by cooler air from above because of atmospheric mixing. Temperature begins rising as soon as the net radiation budget of the surface becomes positive. Temperature continues to rise from sunrise until sometime after solar noon. After this time, mixing of the Earth's surface by convection causes the surface to cool despite the positive addition of radiation and heat energy.

ANNUAL CYCLE OF AIR TEMPERATURE

As the Earth revolves around the Sun, locations on the surface may under go seasonal changes in air temperature because of annual variations in the intensity of net radiation. Variations in net radiation are primarily controlled by changes in the intensity and duration of received solar insolation which are driven by variations in daylength and angle of incidence. The discussion below examines how changes in net radiation can effect *mean monthly temperatures* for the following five locations:

- *Manaus*, Brazil, 3° South latitude.
- *Bulawayo*, Zimbabwe, 20° South latitude.
- *Albuquerque*, USA, 35° North latitude.
- *London*, England, 52° North latitude.
- *Fairbanks*, USA, 65° North latitude.

Manaus, Brazil - 3° South, 60° West:

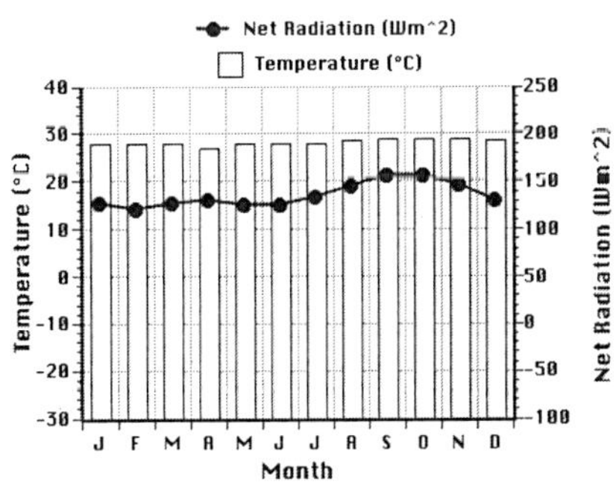

Fig. Monthly variations in net radiation and average monthly temperature for Manaus, Brazil.

At *Manaus*, values of monthly net radiation average about 135 Watts per square meter. Monthly variation in net radiation is only about 35 Watts over the entire year. Two peaks in net radiation are visible on the graph. Both of these peaks occur during the equinoxes when the height of the Sun above the horizon is at its maximum (90° above the horizon). Minimum values of net radiation correspond to the time of the year when the Sun reaches its minimum height of only 66.5° above the horizon at solar noon. Because of the consistent nature of net radiation, mean monthly air temperature only varies by 2° Celsius over the entire year.

Bulawayo, Zimbabwe - 20° South, 29° East

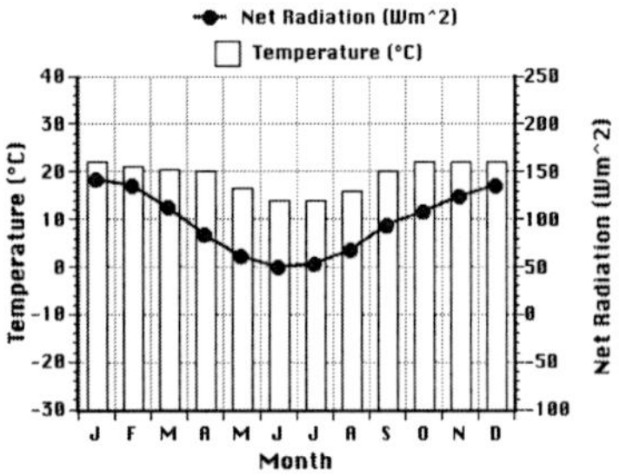

Fig. Monthly variations in net radiation and average monthly temperature for Bulawayo, Zimbabwe.

Net radiation at *Bulawayo* has a single peak and trough over the one year period graphed. This pattern is primarily controlled by variations in the intensity and duration of incoming solar insolation. During the December solstice, the Sun reaches its highest altitude above the horizon and daylength is at a maximum (13 hours and 12 minutes). The lowest values of net radiation occur around the June solstice when the Sun reaches its lowest altitude above the horizon and daylength is at a minimum (10 hours and 48 minutes) in the Southern Hemisphere. Monthly temperature variations follow the monthly change in net radiation. Net radiation represents energy available to do work. When received at the Earth's surface much of this energy is used to create sensible heat.

Albuquerque, USA - 35° North, 107° West

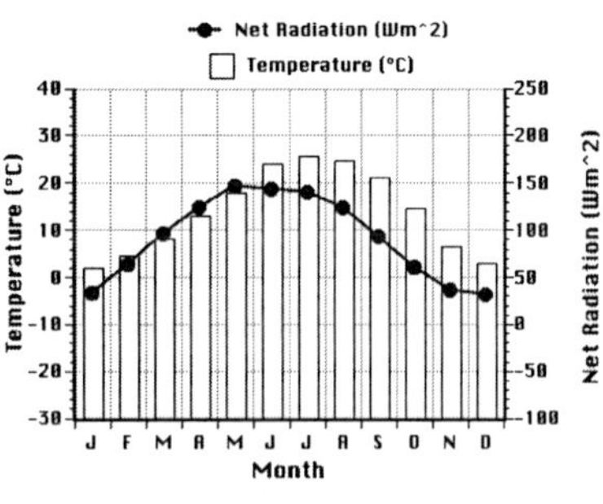

Fig. Monthly variations in net radiation and average monthly temperature for Albuquerque, USA.

At *Albuquerque*, maximum net radiation occurs in May. The timing of this peak roughly coincides with the June solstice when daylengths are at their longest and solar heights are their greatest. However, monthly temperature variations do not mirror the changes in net radiation exactly. Peak monthly temperatures occur about two months after the net radiation maximum. This lag is probably caused by the delayed movement of stored heat energy in the ground into the atmosphere. Minimum monthly temperatures do coincide with the lowest values of net radiation which occur during the December solstice.

London, England - 52° North, 1° East

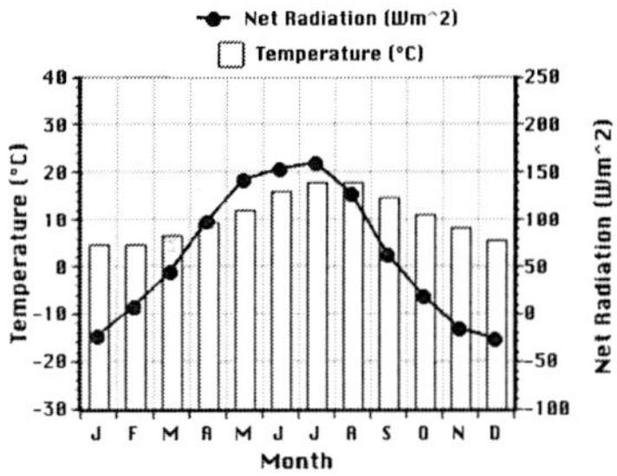

Fig. Monthly variations in net radiation and average monthly temperature for London, England.

The annual patterns of net radiation and mean monthly temperature for *London* are quite similar to those already described for Albuquerque. London does, however, experience a greater annual variation in net radiation. This greater variation can be explained by the effect increasing latitude has on annual variations of insolation. During the winter months, outgoing longwave radiation actually exceeds incoming insolation producing negative net radiation values. This was not seen in Albuquerque. The variation in monthly mean temperature is also less extreme in London when compared to Albuquerque. Intuitively, one would expect London to have a greater annual change in temperature because of the greater variation in net radiation over the year. However, London's climate is moderated by the frequent addition of latent heat energy from seasonal precipitation.

Fairbanks, USA - 65° North, 148° West

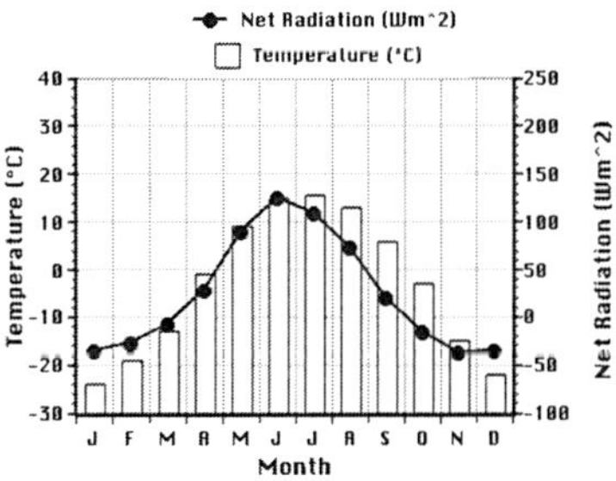

Fig. Monthly variations in net radiation and average monthly temperature for Fairbanks, USA.

Of the five locations examined, *Fairbanks* has the greatest variations in mean monthly temperature. Fairbanks is also the coldest of the climates examined. This is primarily due to the fact that during six months of the year net radiation is negative because outgoing longwave radiation exceeds incoming insolation. Fairbanks also receives the least cumulative amount of net radiation over the entire year. Mean month temperature is at its maximum in July which is one month ahead of the peak in net radiation.

GLOBAL SURFACE TEMPERATURE DISTRIBUTION

If the Earth was a homogeneous body without the present land/ocean distribution, its temperature distribution would be strictly latitudinal. However, the Earth is more complex than this being composed of a mosaic of land and water. This mosaic causes latitudinal zonation of temperature to be disrupted spatially.

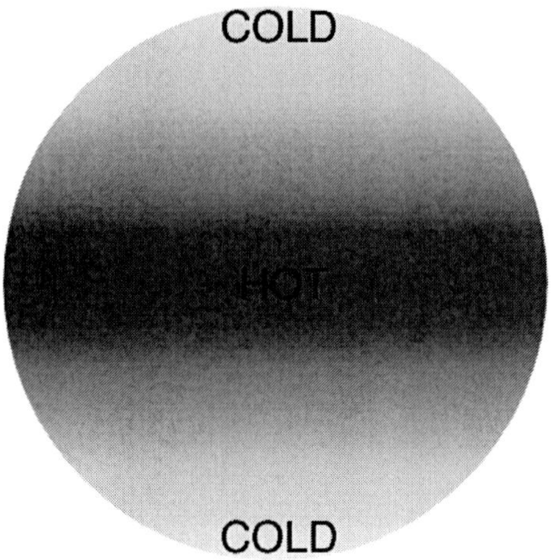

Fig. Simple latitudinal zonation of temperature.

The following two factors are important in influencing the distribution of temperature on the Earth's surface:

- The latitude of the location determines how much solar radiation is received. Latitude influences the angle of incidence and duration of daylength.
- Surface properties - surfaces with high albedo absorb less incident radiation. In general, land absorbs less insolation that water because of its lighter Colour. Also, even if two surfaces have the same albedo, a surface's specific heat determines the amount of heat energy required for a specific rise in temperature per unit mass. The specific heat of water is some five times greater than that of rock and the land surface. As a result, water requires the input of large amounts of energy to cause a rise in its temperature.

Table. Specific Heat of Various Substances.

Substance	Specific Heat
Water	1.00
Air	0.24
Granite	0.19
Sand	0.19
Iron	0.11

Mainly because of specific heat, land surfaces behave quite differently from water surfaces. In general, the surface of any extensive deep body of water heats more slowly and cools more slowly than the surface of a large land body. Other factors influencing the way land and water surfaces heat and cool include:

- Solar radiation warms an extensive layer in water, on land just the immediate surface is heated.
- Water is easily mixed by the process of convection.
- Evaporation of water removes energy from water's surface.

The following images illustrate the Earth's temperature distribution patterns for an average January and July based on 39 years of data. Note that the spatial variations of temperature on these figures is mostly latitudinal. However, the horizontal banding of isotherms is somewhat upset by the fact that water heats up more slowly in the summer and cools down more slowly in the winter when compared to land surfaces. During January, much of the terrestrial areas of the Northern Hemisphere are below freezing. Some notable Northern Hemisphere cold-spots include the area around Baffin Island Canada, Greenland, Siberia, and the Plateau of Tibet. Temperatures over oceans tend to be hotter because of the water's ability to hold heat energy.

In the Southern Hemisphere, temperatures over the major landmasses are generally greater than 20° Celsius with localized hot-spots in west-central Australia, the Kalahari Desert in Africa, and the plains of Bolivia, Paraguay, and Argentina. Subtropical oceans are often warmer than landmass areas near the equator. At this latitude, land areas receive less incoming solar radiation because of the daily convective development of cumulus and cumulonimbus clouds. In the mid-latitudes, oceans are often cooler than landmass areas at similar latitudes. Terrestrial areas are warmer because of the rapid heating of land surfaces under frequently clear skies. Antarctica remains cold and below zero degrees Celsius due to the presence of permanent glacial ice which reflects much of the solar radiation received back to space.

In July, the Northern Hemisphere is experiencing its summer season because the North Pole is now tilted towards the Sun. Some conspicuous hot-spots include the south-central United States, Arizona and northwest Mexico, northern Africa, the Middle East, India, Pakistan, and Afghanistan. Temperatures over oceans tend to be relatively cooler because of the land's ability to heat quickly. Two terrestrial areas of cooler temperatures include

Greenland and the Plateau of Tibet. In these regions, most of the incoming solar radiation is sent back to space because of the presence of reflective ice and snow. In the Southern Hemisphere, temperatures over the major landmasses are generally cooler than ocean surfaces at the same latitude. Antarctica is bitterly cold because it is experiencing total darkness. Note that Antarctica is much colder than the Arctic was during its winter season. The Arctic consists mainly of ocean. During the summer, this surface is able to absorb considerable quantities of sunlight which is then converted into heat energy. The heat stored in the ocean is carried over into the winter season. Antarctica has a surface composed primarily of snow and ice. This surface absorbs only a small amount of the solar radiation during the summer. So it never really heats up. As a result, the amount of heat energy stored into the winter season is minimal.

FORCES ACTING TO CREATE WIND

Wind can be defined simply as air in motion. This motion can be in any direction, but in most cases the horizontal component of wind flow greatly exceeds the flow that occurs vertically. The speed of wind varies from absolute calm to speeds as high as 380 kilometers per hour. In 1894, strong winds in Nebraska pushed six fully loaded coal cars over 160 kilometers in just over three hours. Over short periods of time surface winds can be quite variable.

Wind develops as a result of spatial differences in atmospheric pressure. Generally, these differences occur because of uneven absorption of solar radiation at the Earth's surface. Wind speed tends to be at its greatest during the daytime when the greatest spatial extremes in atmospheric temperature and pressure exist.

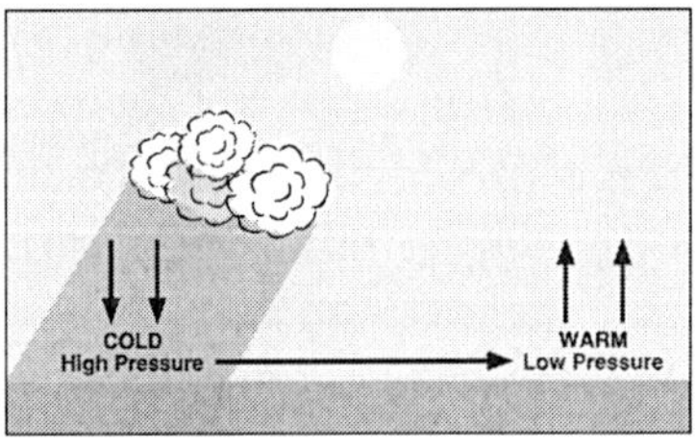

Fig. Formation of wind as a result of localized temperature differences.

Wind is often described by two characteristics: wind speed and wind direction. Wind speed is the velocity attained by a mass of air traveling horizontally through the atmosphere. Wind speed is often measured with an anemometer in kilometers per hour (kmph), miles per hour (mph), knots, or meters per second (mps). Wind direction is measured as the direction from where a wind comes from. For example, a southerly wind comes from the south and blows to the north. Direction is measured by an instrument called a wind vane. Both of these instruments are positioned in the atmospheric environment at a standard distance of 10 meters above the ground surface.

Wind speed can also be measured without the aid of instruments using the Beaufort wind scale. This descriptive scale was originally developed by Admiral Beaufort of the British Navy in the first decade of the 17th century. The purpose for this system was to allow mariners to determine wind speed from simple observations. The Beaufort system has undergone several modifications to standardize its measurement scale and to allow for its use on land. Users of this scale look for specific effects of the wind on the environment to determine speed.

Fig. Meteorological instruments used to measure wind speed and direction. Wind speed is commonly measured with an anemometer. An anemometer consists of three open cups attached to a rotating spindle. The speed of roation is then converted into a measurement of wind speed. Wind direction is measured with a wind vane. On the photograph above, the wind vane instrument has a bullet shaped nose attached to a finned tail by a metal bar.

Table. Beaufort wind speed scale.

Beaufort Code	Speed Miles per Hour	Speed Kilometers per Hour	Description	Effects on the Environment
0	< 1	< 1	calm	smoke rises vertically
1	2 - 3	1 - 5	light air	smoke drifts slowly
2	4 - 7	6 - 11	light breeze	leaves rustle, wind can be felt, wind vanes move
3	8 - 12	12 - 19	gentle breeze	leaves and twigs on trees move
4	13 - 18	20 - 29	moderate breeze	small tree branches move, dust is picked up from the ground surface
5	19 - 24	30 - 38	fresh breeze	small trees move
6	25 31	39 - 51	strong breeze	large branches move, telephone and power overhead wires whistle
7	32 - 38	51 - 61	near gale	trees move, difficult to walk in the wind
8	39 - 46	62 - 74	gale	twigs break off from trees
9	47 - 54	75 - 86	strong gale	branches break off from trees, shingles blown off roofs
10	55 - 63	87 - 101	whole gale	trees become uprooted, structural damage on buildings
11	64 - 74	102 - 120	storm	widespread damage to buildings and trees
12	> 75	> 120	hurricane	severe damage to buildings and trees

Winds are named according to the compass direction of their source. Thus, a wind from the north blowing Towards the south is called a northerly wind. Figure 7n-3 describes the sixteen principal bearings of wind direction. Most meterological observations report wind direction using one of these sixteen bearings.

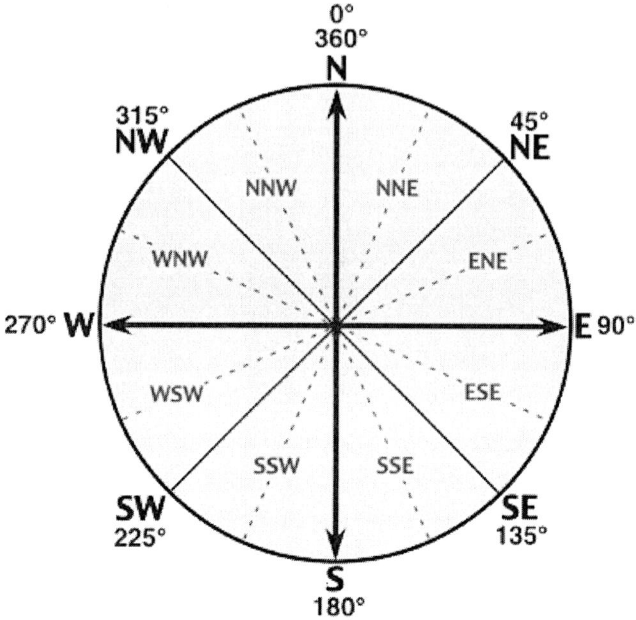

Fig. Wind compass describing the sixteen principal bearings used to measure wind direction. This compass is based on the 360 degrees found in a circle.

Horizontally, at the Earth's surface wind always blows from areas of high pressure to areas of low pressure (vertically, winds move from areas of low pressure to areas of high pressure), usually at speeds determined by the rate of air pressure change between pressure centers. This situation is comparable to someone skiing down a hill. The skier will of course move from the top of the hill to the bottom of the hill, with the speed of their descent controlled by the gradient or steepness of the slope. Likewise, wind speed is a function of the steepness or gradient of atmospheric air pressure found between high and low pressure systems. When expressed scientifically, pressure change over a unit distance is called pressure gradient force, and the greater this force the faster the winds will blow.On weather maps, pressure is indicated by drawing isolines of pressure, called isobars, at regular 4 millibar intervals (*e.g.*, 996 mb, 1000 mb, 1004 mb, etc.). If the isobars are closely spaced, we can expect the pressure gradient force to be great, and wind speed to be high. In areas where the isobars are spaced widely apart, the pressure gradient is low and light winds normally exist. High speed winds develop in areas where isobars are closer.

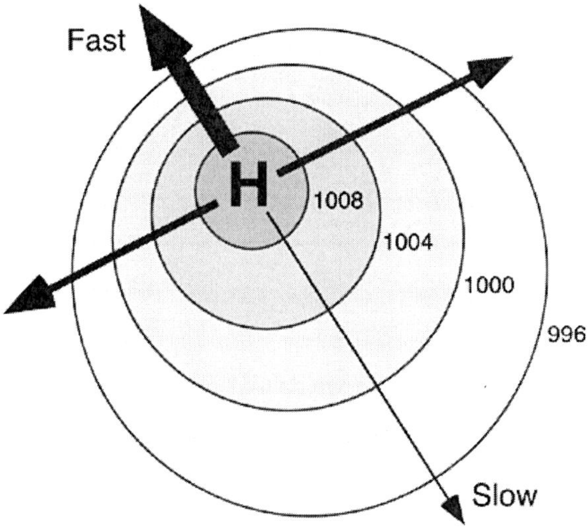

Fig. Association between wind speed and distance between isobars. In the illustration above thicker arrows represent relatively faster winds.

DRIVING FORCES

To better understand wind we must recognize that it is the result of a limited number of accelerating and decelerating forces, and that the action of these forces is controlled by specific fundamental natural laws. Sir Isaac Newton formulated these laws as several laws of motion. The first law suggests that an object that is stationary will remain stationary, and an object in motion will stay in motion as long as no opposing force is put on the object. As a result of this law, a puck sent in flight from a blade of a hockey stick will remain in motion until friction slows it down or the goalie makes a save. This law also suggests that once in motion an object's path should be straight.

Newton's second law of motion suggests that the force put on an object equals its mass multiplied by the acceleration produced. The term force in this law refers to the total or net effect of all the forces acting on an object. Mathematically, this law is written as:

- Force = Mass x Acceleration, or
- Acceleration = Force/Mass

From this natural law of motion we can see that the acceleration of an object is directly proportional to the net force pushing or pulling that body and inversely proportional to the mass of the body. Thus, the greater the force created by the movement of a hockey player's stick the faster the puck will travel. This law also suggests that if the player used a larger (more massive) puck more force would have to be applied to it to get it to travel as fast as a puck with less mass. In the previous lecture we briefly examined one of the forces, pressure gradient force, acting on wind. Let us return to this force and examine it in greater detail and in relation to Newton's laws of motion. We will also examine the effects of three other forces that act on air in motion.

Pressure gradient force is the primary force influencing the formation of wind from local to global scales. This force is determined by the spatial pattern of atmospheric pressure at any given moment in time. Figure 7n-5 illustrates two different pressure gradient scenarios and their relative effect on wind speed.

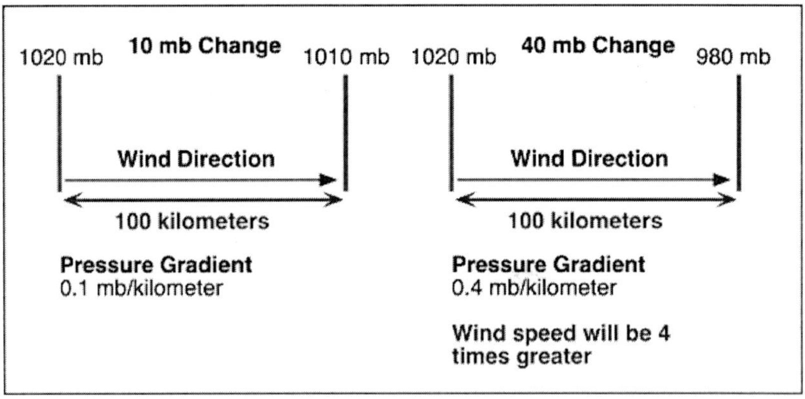

Figure: Effect of pressure gradient on wind speed.

The two diagrams display the relative relationship between pressure gradient and wind speed. This relationship is linear and positive. As a result, quadrupling the pressure gradient increases wind speed by a factor of four. This is what we would expect according to Newton's second law of motion, assuming the mass of the wind is unchanged. We can also describe pressure gradient acceleration mathematically with the following equation:

$$F(m/s^2) = \left| \frac{1}{D} \bullet \left(\frac{P_1 - P_2}{n} \right) \right|$$

where:

- D = density of air (average density of surface air is 1.29 kilograms per cubic meter)
- P2 = pressure at point 2 in Newtons/m^2 (N m^{-2})
- P1 = pressure at point 1 in Newtons/m^2 (N m^{-2})
- n = distance between the two points in meters

From this equation we can determine wind acceleration between two points in meters per second squared by knowing three variables: the density of the moving air; the change in pressure between the points of interest in newtons; and the distance between the two points in meters. For example, to determine the wind speed between two points for moving air with a density of 1.29 kilograms per cubic meter, a pressure difference of 400 Newtons/m^2, and a distance of 300,000 meters, the following calculations would be performed:

$$F(m/s^2) = \left| \frac{1}{1.29} \bullet \left(\frac{400}{300,000} \right) \right| = 0.00103 m/s^2$$

From the calculated value of acceleration we can determine wind speed, V, from the formula:

- V = V0 + Ft

Where V0 is the initial velocity of the wind and t is the time during which F is applied.

The rotation of the Earth creates another force, termed Coriolis force, which acts upon wind and other objects in motion in very predictable ways. According to Newton's first law of motion, air will remain moving in a straight line unless it is influenced by an unbalancing force. The consequence of Coriolis force opposing pressure gradient acceleration is that the moving air changes direction. Instead of wind blowing directly from high to low pressure, the rotation of the Earth causes wind to be deflected off course. In the Northern Hemisphere, wind is deflected to the right of its path, while in the Southern Hemisphere it is deflected to the left. The magnitude of the Coriolis force varies with the velocity and the latitude of the object. Coriolis force is absent at the equator, and its strength increases as one approaches either pole. Furthermore, an increase in wind speed also results in a stronger Coriolis force, and thus in greater deflection of the wind. Coriolis force only acts on air when it has been sent into motion by pressure gradient force. Finally, Coriolis force only influences wind direction and never wind speed.

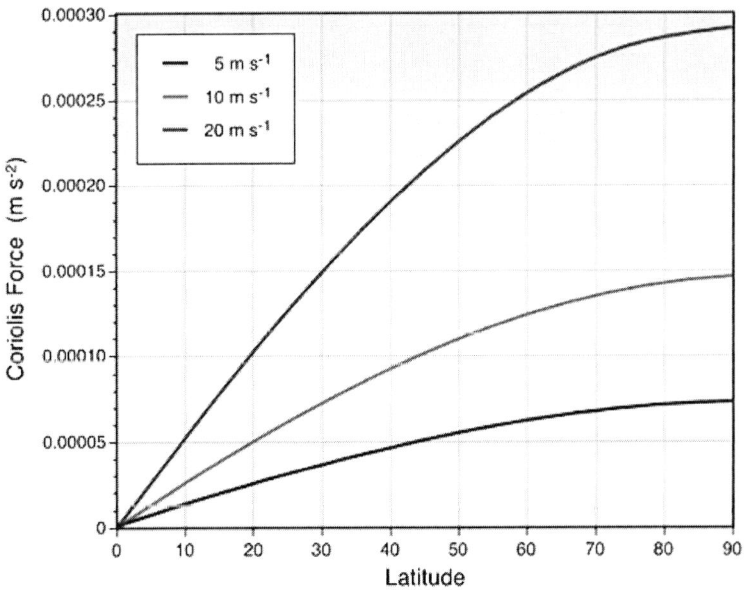

Fig. The strength of Coriolis force is influenced by latitude and the speed of the moving object.

Centripetal acceleration is the third force that can act on moving air. It acts only on air that is flowing around centers of circulation. Centripetal acceleration is also another force that can influence the direction of wind. Centripetal acceleration creates a force directed at right angles to the flow of the wind and inwards towards the centers of rotation (*e.g.*, low and high pressure centers). This force produces a circular pattern of flow around centers of high and low pressure. Centripetal acceleration is much more important for circulations smaller than the mid-latitude cyclone.

The last force that can influence moving air is frictional deceleration. Friction can exert an influence on wind only after the air is in motion. Frictional drag acts in a direction opposite to the path of motion causing the moving air to decelerate. Frictional effects are limited to the lower one kilometer above the Earth's surface.

GEOSTROPHIC WIND

Air under the influence of both the pressure gradient force and Coriolis force tends to move parallel to isobars in conditions where friction is low (1000 meters above the surface of the Earth) and isobars are straight. Winds of this type are usually called geostrophic winds. Geostrophic winds come about because pressure gradient force and Coriolis force come into balance after the air begins to move.

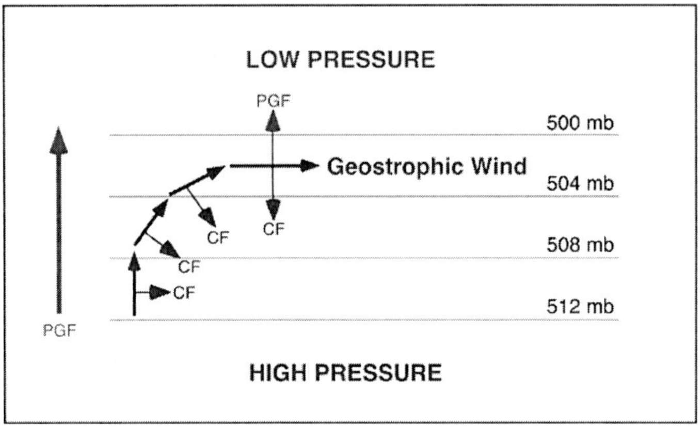

Fig. A geostrophic wind flows parallel to the isobars. In this model of wind flow in the Northern Hemisphere, wind begins as a flow of air perpendicular to the isobars (measured in millibars) under the primary influence of the pressure gradient force (PGF). As the movement begins, the Coriolis force (CF) begins to influence the moving air causing it to deflect to the right of its path. This deflection continues until the pressure gradient force and Coriolis force are opposite and in balance with each other.

Figure 7n-7 models air flow in the Northern Hemisphere. In the Southern Hemisphere, Coriolis acceleration acts on moving air by deflecting it to the left instead of the right. Finally, Buy Ballot's Law states that when you stand with your back to a geostrophic wind in the Northern Hemisphere the center of low pressure will be to your left and the high pressure to your right. The opposite is true for the Southern Hemisphere.

GRADIENT WIND

Wind above the Earth's surface does not always travel in straight lines. In many cases winds flow around the curved isobars of a high (anticyclone) or low (cyclone) pressure center. A wind that blows around curved isobars above the level of friction is called a gradient wind. Gradient winds are slightly more complex than geostrophic winds because they include the action of yet another physical force.

This force is known as centripetal force and it is always directed Towards the center of rotation. The following figure describes the forces that produce gradient winds around high and low pressure centers. Around a low, the gradient wind consists of the pressure gradient force and centripetal force acting Towards the center of rotation, while Coriolis force acts away from the center of the low. In a high pressure center, the Coriolis and centripetal forces are directed Towards the center of the high, while the pressure gradient force is directed outward.

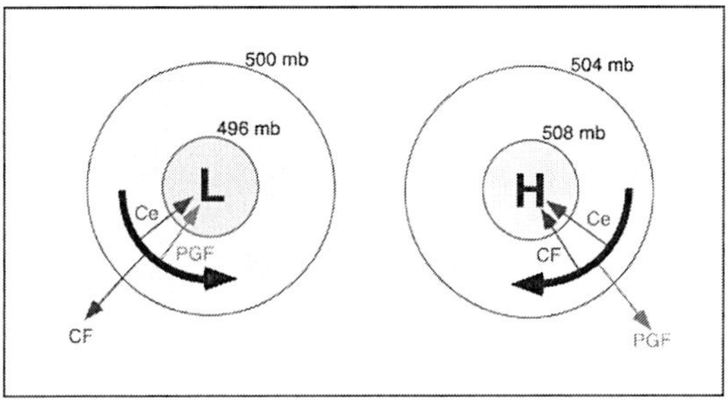

Fig. The balance of forces that create a gradient wind in the Northern Hemisphere (PGF = pressure gradient force; CF = Coriolis force; Ce = centripetal force). In this diagram, CF = Ce + PGF for the low, and PGF = CF + Ce for the high.

FRICTION LAYER WIND

Surface winds on a weather map do not blow exactly parallel to the isobars as in geostrophic and gradient winds. Instead, surface winds tend to cross the isobars at an angle varying from 10 to 45°. Close to the Earth's surface, friction reduces the wind speed, which in turn reduces the Coriolis force. As a result, the reduced Coriolis force no longer balances the pressure gradient force, and the wind blows across the isobars Towards or away from the pressure center. The pressure gradient force is now balanced by the sum of the frictional force and the Coriolis force.

Thus, we find surface winds blowing counterclockwise and inward into a surface low, and clockwise and out of a surface high in the Northern Hemisphere. In the Southern Hemisphere, the Coriolis force acts to the left rather than the right. This causes the winds of the Southern Hemisphere to blow clockwise and inward around surface lows, and counterclockwise and outward around surface highs.

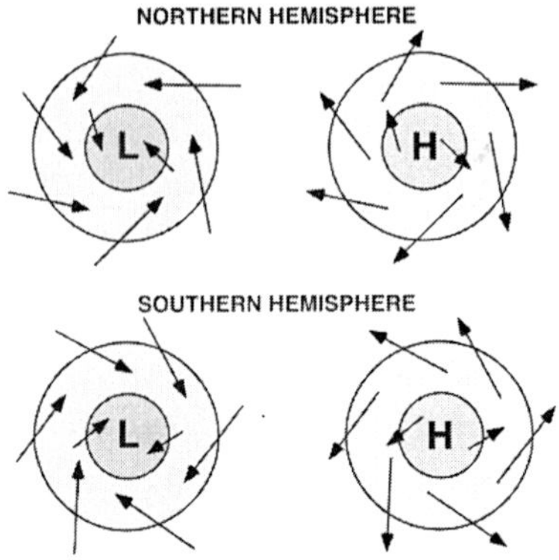

Fig. Circulation patterns of high and low pressure systems in the North and South Hemisphere.

LOCAL AND REGIONAL WIND SYSTEMS

THERMAL CIRCULATIONS

As discussed earlier, winds blow because of differences in atmospheric pressure. Pressure gradients may develop on a local to a global scale because of differences in the heating and cooling of the Earth's surface. Heating and cooling cycles that develop daily or annually can create several common local or regional thermal wind systems. The basic circulation system that develops is described in the generic illustrations below.

Fig. Cross-section of the atmosphere with uniform horizontal atmospheric pressure.

In this first diagram, there is no horizontal temperature or pressure gradient and therefore no wind. Atmospheric pressure decreases with altitude as depicted by the drawn isobars (1000 to 980 millibars). In the second diagram, the potential for solar heating is added which creates contrasting surface areas of temperature and atmospheric pressure. The area to the right receives more solar radiation and the air begins to warm from heat energy transferred from the ground through conduction and convection. The vertical distance between the isobars becomes greater as the air rises. To the far left, less radiation is received because of the presence of cloud, and this area becomes relatively cooler than the area to the right. In the upper atmosphere, a pressure gradient begins to form because of the rising air and upward spreading of the isobars. The air then begins to flow in the upper atmosphere from high pressure to low pressure.

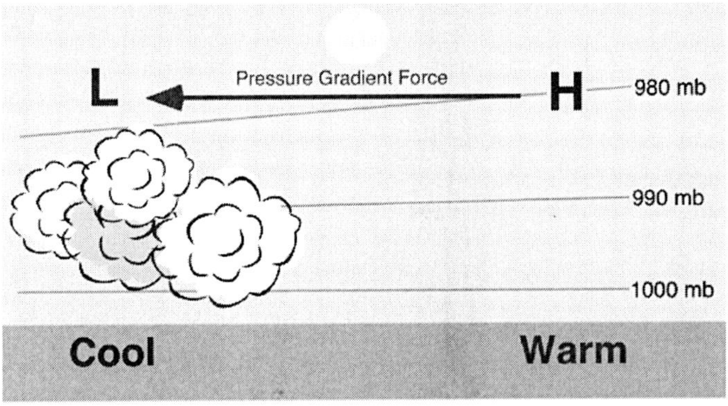

Fig. Development of air flow in the upper atmosphere because of surface heating.

Figure 7o-3 shows the full circulation system in action. Beneath the upper atmosphere high is a thermal low pressure center created from the heating of the ground surface. Below the upper atmosphere low is a thermal high created by the relatively cooler air temperatures and enhanced by the descending air from above. Surface air temperatures are cooler here because of the obstruction of shortwave radiation absorption at the Earth's surface by the cloud. At the surface, the wind blows from the high to the low pressure. Once at the low, the wind rises up to the upper air high pressure system because of thermal buoyancy and outflow in the upper atmosphere. From the upper high, the air then travels to the upper air low, and then back down to the surface high to complete the circulation cell. The circulation cell is a closed system that redistributes air in an equitable manner. It is driven by the greater heating of the surface air in the right of the diagram.

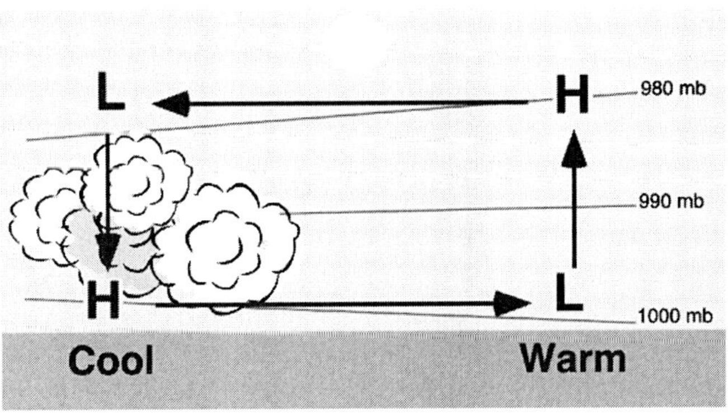

Fig. Development of a closed atmospheric circulation cell because of surface heating.

SEA AND LAND BREEZES

Sea and land breezes are types of thermal circulation systems that develop at the interface of land and ocean. At this interface, the dissimilar heating and cooling characteristics of land and water initiate the development of an atmospheric pressure gradient which causes the air in these areas to flow.

During the daytime land heats up much faster than water as it receives solar radiation from the Sun. The warmer air over the land then begins to expand and rise forming a thermal low.

At the same time, the air over the ocean becomes a cool high because of water's slower rate of heating. Air begins to flow as soon as there is a significant difference in air temperature and pressure across the land to sea gradient.

The development of this pressure gradient causes the heavier cooler air over the ocean to move Towards the land and to replace the air rising in the thermal low. This localized air flow system is called a sea breeze. Sea breeze usually begins in midmorning and reaches its maximum strength in the later afternoon when the greatest temperature and pressure contrasts exist. It dies down at sunset when air temperature and pressure once again become similar across the two surfaces.

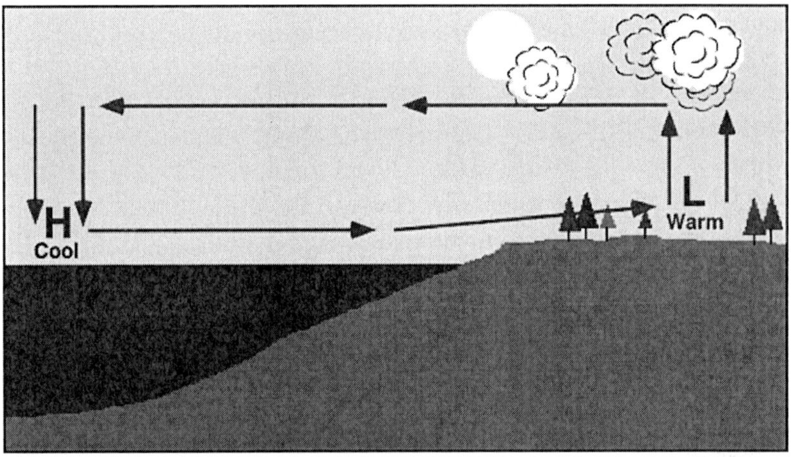

Figure: Daytime development of sea breeze.

At sunset, the land surface stops receiving radiation from the Sun. As night continues the land surface begins losing heat energy at a much faster rate than the water surface.

After a few hours, air temperature and pressure contrasts begin to develop between the land and ocean surfaces. The land surface being cooler than the water becomes a thermal high pressure area. The ocean becomes a warm thermal low. Wind flow now moves from the land to the open ocean. This type of localized air flow is called a land breeze.

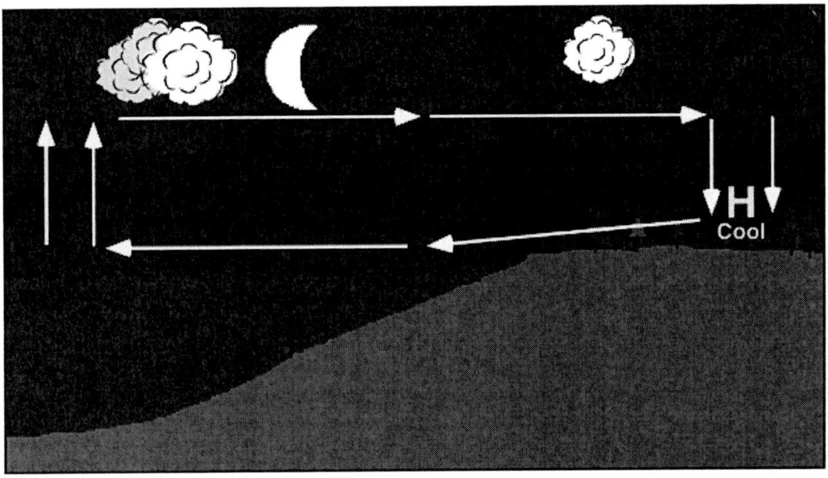

Fig. Nighttime development of land breeze.

MOUNTAIN AND VALLEY BREEZES

Mountain and valley breezes are common in regions with great topographic relief. A valley breeze develops during the day as the Sun heats the land surface and air at the valley bottom and sides. As the air heats it becomes less dense and buoyant and begins to flow gently up the valley sides. Vertical ascent of the air rising along the sides of the mountain is usually limited by the presence of a temperature inversion layer. When the ascending air currents encounter the inversion they are forced to move horizontally and then back down to the valley floor. This creates a self-contained circulation system. If conditions are right, the rising air can condense and form into cumuliform clouds.

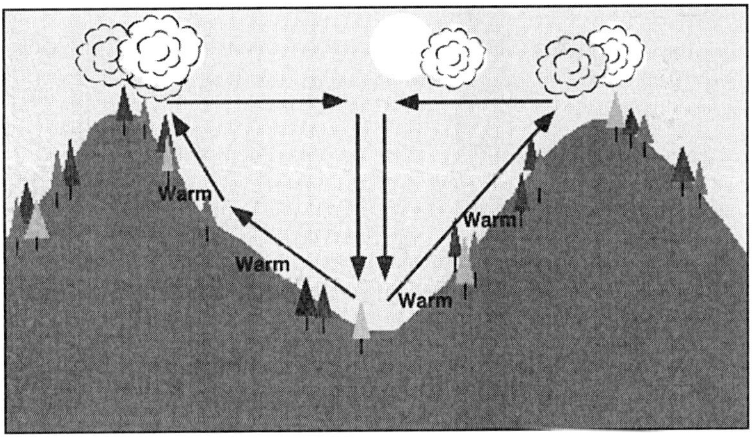

Fig. Daytime development of valley breeze.

During the night, the air along the mountain slopes begins to cool quickly because of longwave radiation loss. As the air cools, it becomes more dense and begins to flow downslope causing a mountain breeze. Convergence of the draining air occurs at the valley floor and forces the air to move vertically upward. The upward movement is usually limited by the presence of a temperature inversion which forces the air to begin moving horizontally. This horizontal movement completes the circulation cell system. In narrowing terrain, mountain winds can accelerate in speed because of the venturi effect. Such winds can attain speeds as high has 150 kilometers per hour.

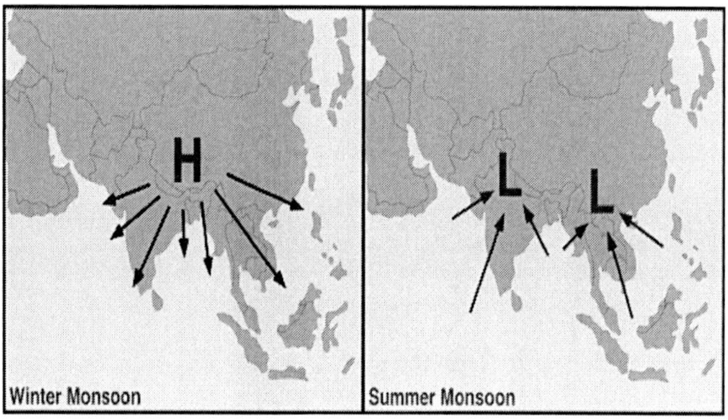

Fig. Nightime development of mountain breeze.

MONSOON WINDS

Monsoons are regional scale wind systems that predictably change direction with the passing of the seasons. Like land/sea breezes, these wind systems are created by the temperature contrasts that exist between the surfaces of land and ocean. However, monsoons are different from land/sea breezes both spatially and temporally. Monsoons occur over distances of thousands of kilometers, and their two dominant patterns of wind flow act over an annual time scale. During the summer, monsoon winds blow from the cooler ocean surfaces onto the warmer continents. In the summer, the continents become much warmer than the oceans because of a number of factors. These factors include:

- Specific heat differences between land and water.
- Greater evaporation over water surfaces.
- Subsurface mixing in ocean basins which redistributes heat energy through a deeper layer.

Precipitation is normally associated with the summer monsoons. Onshore winds blowing inland from the warm ocean are very high in humidity, and slight cooling of these air masses causes condensation and rain. In some cases, this precipitation can be greatly intensified by orographic uplift. Some highland areas in Asia receive more than 10 meters of rain during the summer months.

In the winter, the wind patterns reverse as the ocean surfaces are now warmer. With little solar energy available, the continents begin cooling rapidly as longwave radiation is emitted to space. The ocean surface retains its heat energy longer because of water's high specific heat and subsurface mixing. The winter monsoons bring clear dry weather and winds that flow from land to sea.

Figure 7o-8 illustrates the general wind patterns associated with the winter and summer monsoons in Asia. The Asiatic monsoon is the result of a complex climatic interaction between the distribution of land and water, topography, and tropical and mid-latitudinal circulation. In the summer, a low pressure center forms over northern India and northern Southeast Asia because of higher levels of received solar insolation. Warm moist air is drawn into the thermal lows from air masses over the Indian Ocean. Summer heating also causes the development of a strong latitudinal pressure gradient and the development of an easterly jet stream at an altitude of about 15 kilometers and a latitude of 25° North. The jet stream enhances rainfall in Southeast Asia, in the Arabian Sea, and in South Africa.

When autumn returns to Asia the thermal extremes between land and ocean decrease and the westerlies of the mid-latitudes move in. The easterly jet stream is replaced with strong westerly winds in the upper atmosphere. Subsidence from an upper atmosphere cold low above the Himalayas produces outflow that creates a surface high pressure system that dominates the weather in India and Southeast Asia.Monsoon wind systems also exist in Australia, Africa, South America, and North America.

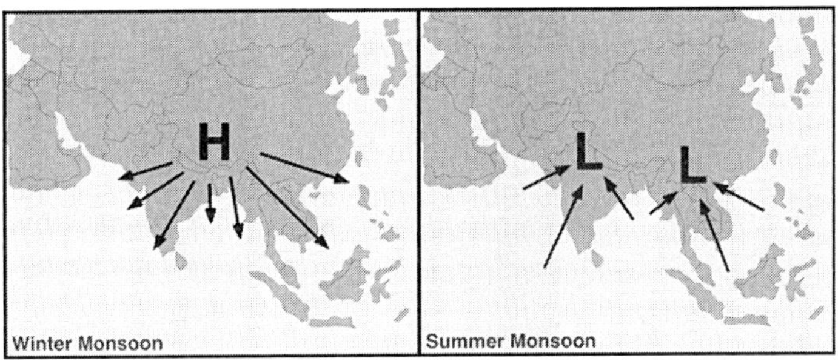

Fig. Winter and summer monsoon wind patterns for southeast Asia.

GLOBAL SCALE CIRCULATION OF THE ATMOSPHERE

SIMPLE MODEL OF GLOBAL CIRCULATION

We can gain an understanding of how global circulation works by developing two simplified graphical models of processes that produce this system. The first model will be founded on the following simplifying assumptions:

- The Earth is not rotating in space.
- The Earth's surface is composed of similar materials.
- The global reception of solar insolation and loss of longwave radiation cause a temperature gradient of hotter air at the equator and colder air at the poles.

Based on these assumptions, air circulation on the Earth should approximate the patterns shown on Figure 7p-1. In this illustration, each hemisphere contains one three-dimensional circulation cell.

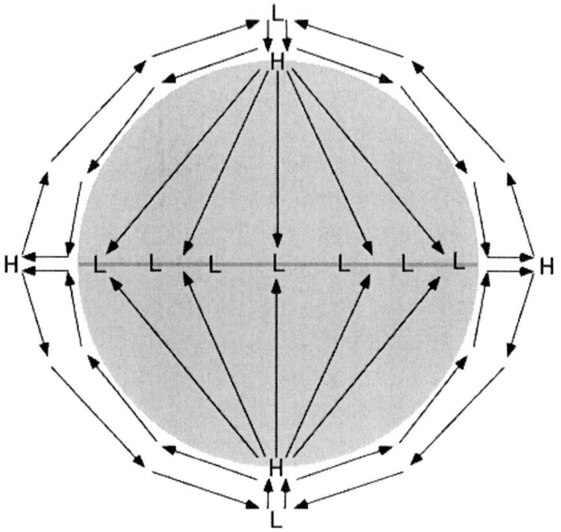

Fig. Simplified one-cell global air circulation patterns.

As described in the diagram above, surface air flow is from the poles to the equator. When the air reaches the equator, it is lifted vertically by the processes of convection and convergence. When it reaches the top of the troposphere, it begins to flow once again horizontally. However, the direction of flow is now from the equator to the poles. At the poles, the air in the upper atmosphere then descends to the Earth's surface to complete the cycle of flow.

THREE CELL MODEL OF GLOBAL CIRCULATION

If we eliminate the first assumption, the pattern of flow described in the model above would be altered, and the mesoscale flow of the atmosphere would more closely approximate the actual global circulation on the Earth. Planetary rotation would cause the development of three circulation cells in each hemisphere rather than one. These three circulation cells are known as the: Hadley cell; Ferrel cell; and Polar cell.

In the new model, the equator still remains the warmest location on the Earth. This area of greater heat acts as a zone of thermal lows known as the intertropical convergence zone (ITCZ). The Intertropical Convergence Zone draws in surface air from the subtropics. When this subtropical air reaches the equator, it rises into the upper atmosphere because of convergence and convection. It attains a maximum vertical altitude of about 14 kilometers (top of the troposphere), and then begins flowing horizontally to the North and South

Poles. Coriolis force causes the deflection of this moving air in the upper atmosphere, and by about 30° of latitude the air begins to flow zonally from west to east. This zonal flow is known as the subtropical jet stream. The zonal flow also causes the accumulation of air in the upper atmosphere as it is no longer flowing meridionally.

To compensate for this accumulation, some of the air in the upper atmosphere sinks back to the surface creating the subtropical high pressure zone. From this zone, the surface air travels in two directions

. A portion of the air moves back Towards the equator completing the circulation system known as the Hadley cell. This moving air is also deflected by the Coriolis effect to create the Northeast Trades (right deflection) and Southeast

Trades (left deflection). The surface air moving towards the poles from the subtropical high zone is also deflected by Coriolis acceleration producing the Westerlies.

Between the latitudes of 30 to 60° North and South, upper air winds blow generally towards the poles. Once again, Coriolis force deflects this wind to cause it to flow west to east forming the polar jet stream at roughly 60° North and South.

On the Earth's surface at 60° North and South latitude, the subtropical Westerlies collide with cold air traveling from the poles. This collision results in frontal uplift and the creation of the subpolar lows or mid-latitude cyclones. A small portion of this lifted air is sent back into the Ferrel cell after it reaches the top of the troposphere.

Most of this lifted air is directed to the polar vortex where it moves downward to create the polar high.

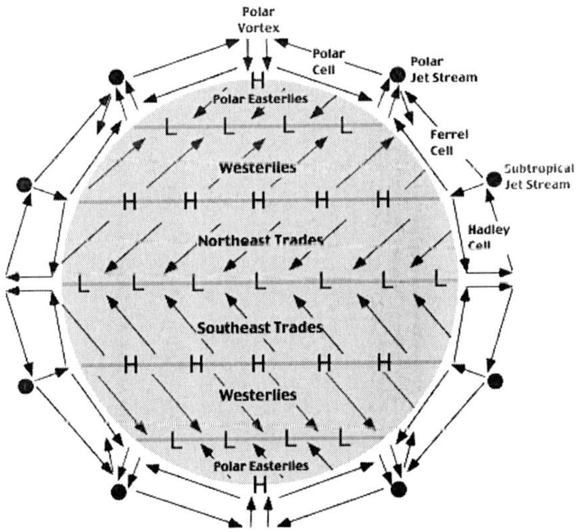

Fig. Simplified global three-cell surface and upper air circulation patterns.

The subtropical high pressure zone does not form a uniform area of high pressure stretching around the world in reality. Instead, the system consists of several localized anticyclonic cells of high pressure.

These systems are located roughly at about 20 to 30° of latitude and are labeled with the letter H on Figures 7p-4 and 7p-5. The subtropical high pressure systems develop because of the presence of descending air currents from the Hadley cell.

These systems intensify over the ocean during the summer or high Sun season. During this season, the air over the ocean bodies remains relatively cool because of the slower heating of water relative to land surfaces. Over land, intensification takes place in the winter months. At this time, land cools off quickly, relative to ocean, forming large cold continental air masses.

The subpolar lows form a continuous zone of low pressure in the Southern Hemisphere at a latitude of between 50 and 70°. The intensity of the subpolar lows varies with season. This zone is most intense during Southern Hemisphere summer. At this time, greater differences in temperature exist between air masses found either side of this zone.

North of subpolar low belt, summer heating warms subtropical air masses. South of the zone, the ice covered surface of Antarctica reflects much of the incoming solar radiation back to space. As a consequence, air masses above Antarctica remain cold because very little heating of the ground surface takes place. The meeting of the warm subtropical and cold polar air masses at the subpolar low zone enhances frontal uplift and the formation of inten se low pressure systems.

In the Northern Hemisphere, the subpolar lows do not form a continuous belt circling the globe. Instead, they exist as localized cyclonic centers of low pressure. In the Northern Hemisphere winter, these pressure centers are intense and located over the oceans just to the south of Greenland and the Aleutian Islands.

These areas of low pressure are responsible for spawning many mid-latitude cyclones. The development of the subpolar lows in summer only occurs weakly unlike the Southern Hemisphere. The reason for this phenomenon is that considerable heating of the Earth's surface occurs from 60 to 90° North. As a result, cold polar air masses generally do not form.

UPPER AIR WINDS AND THE JET STREAMS

Winds at the top of the troposphere are generally poleward and westerly in direction. Figure 7q-1 describes these upper air westerlies along with some other associated weather features. Three zones of westerlies can be seen in each hemisphere on this illustration. Each zone is associated with either the Hadley, Ferrel, or Polar circulation cell.

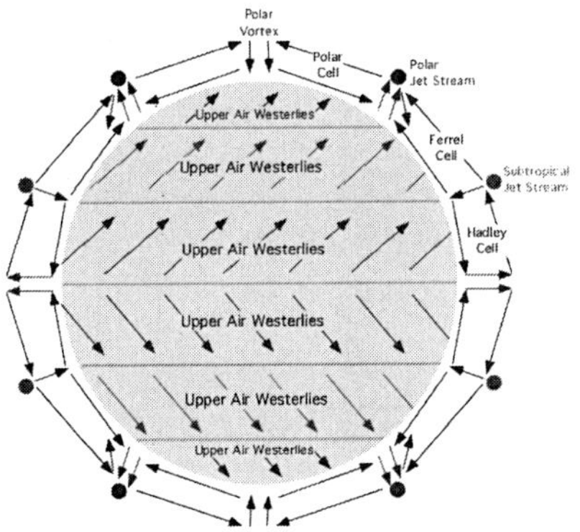

Fig. Simplified global three-cell upper air circulation patterns.

The polar jet stream is formed by the deflection of upper air winds by coriolis acceleration. It resembles a stream of water moving west to east and has an altitude of about 10 kilometers. Its air flow is intensified by the strong temperature and pressure gradient that develops when cold air from the poles meets warm air from the tropics. Wind velocity is highest in the core of the polar jet stream where speeds can be as high as 300 kilometers per hour. The jet stream core is surrounded by slower moving air that has an average velocity of 130 kilometers per hour in winter and 65 kilometers per hour in summer.

Associated with the polar jet stream is the polar front. The polar front represents the zone where warm air from the subtropics and cold air from the poles meet. At this zone, massive exchanges of energy occur in the form of storms known as the mid-latitude cyclones. The shape and position of waves in the polar jet stream determine the location and the intensity of the mid-latitude cyclones. In general, mid-latitude cyclones form beneath polar jet stream troughs. The following satellite image taken from above the South Pole, shows a number of mid-latitude cyclones circling Antarctica. Each mid-latidude cyclone wave is defined by the cloud development associated with frontal uplift.

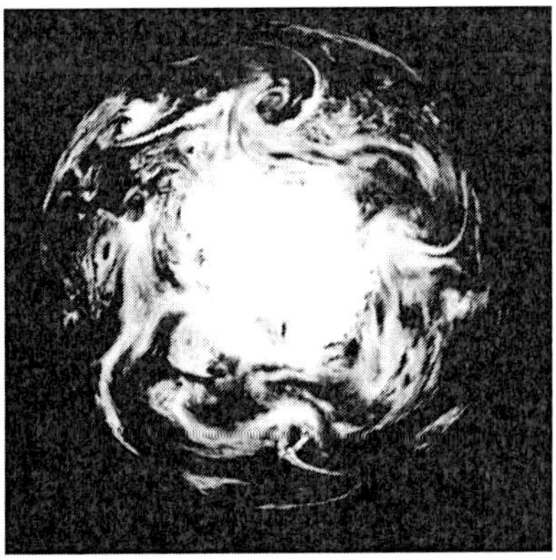

Fig. Satellite view of the atmospheric circulation at the South Pole.

The subtropical jet stream is located approximately 13 kilometers above the subtropical high pressure zone. The reason for its formation is similar to the polar jet stream. However, the subtropical jet stream is weaker. Its slower wind speeds are the result of a weaker latitudinal temperature and pressure gradient.

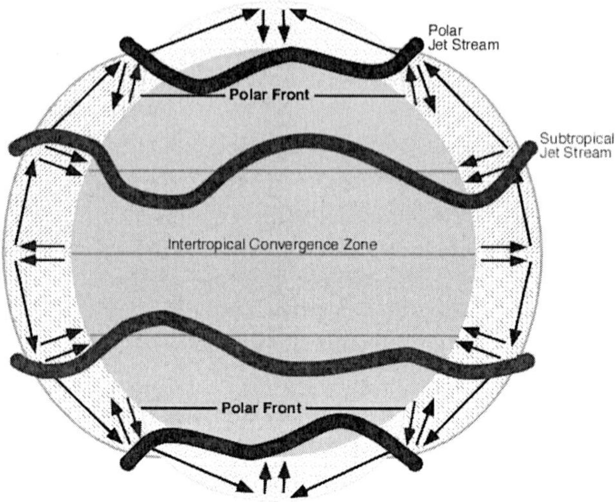

Fig. Polar and subtropical jet streams.

AIR MASSES AND FRONTAL TRANSITIONAL ZONES

An air mass is a large body of air of relatively similar temperature and humidity characteristics covering thousands of square kilometers. Typically, air masses are classified according to the characteristics of their source region or area of formation.

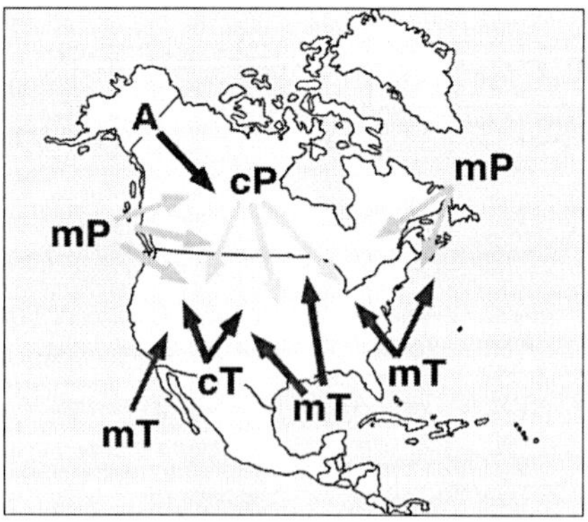

Fig. Source sites and movement patterns for North America's major air masses.

A source region can have one of four temperature attributes: equatorial, tropical, polar or arctic. Air masses are also classified as being either continental or maritime in terms of moisture characteristics. Combining these two categories, several possibilities are commonly found associated with North America: maritime polar (mP), continental polar (cP), maritime tropical (mT), continental tropical (cT), and continental arctic (A). The following diagram describes the source regions and common patterns of movement for the various types of air masses associated with North America.

Frequently, two air masses, especially in the middle latitudes, develop a sharp boundary or interface, where the temperature difference between them becomes intensified. Such an area of intensification is called a frontal zone or a front. The boundary between the warm and cold air masses always slopes upwards over the cold air. This is due to the fact that cold air is much denser than warm air. The sloping of warm air over the cold air leads to a forced uplifting (frontal lifting) of the warm air if one air mass is moving Towards the other. In turn, this uplifting causes condensation to occur and the possibility of precipitation along the frontal boundary.

Frontal zones where the air masses are not moving against each other are called stationary fronts. In transitional areas where there is some air mass movement, cold or warm fronts can develop. Figure 7r-2 illustrates a vertical cross-section of a cold front. A cold front is the transition zone in the atmosphere where an advancing cold, dry stable air mass displaces a warm, moist unstable subtropical air mass. On a weather map, the cold front is drawn as a solid blue line with triangles. The position of the triangles shows the direction of frontal movement. Cold fronts move between 15 to 50 kilometers per hour in a southeast to east direction. The formation of clouds and precipitation at the frontal zone is caused by frontal lifting. High altitude cirrus clouds are found well in advance of the front. Above the surface location of the cold front, high altitude cirrostratus and middle altitude altocumulus are common. Precipitation is normally found just behind the front where frontal lifting has caused the development of towering cumulus and cumulonimbus clouds. Table 7r-1 describes some of the weather conditions associated with a cold front.

Table. Weather conditions associated with a cold front.

Weather Phenomenon	Prior to the Passing of the Front	Contact with the Front	After the Passing of the Front
Temperature	Warm	Cooling suddenly	Cold and getting colder
Atmospheric Pressure	Decreasing steadily	Leveling off then increasing	Increasing steadily
Winds	South to southeast	Variable and gusty	West to northwest
Precipitation	Showers	Heavy rain or snow, hail sometimes	Showers then clearing
Clouds	Cirrus and cirrostratus changing later to cumulus and cumulonimbus	Cumulus and cumulonimbus	Cumulus

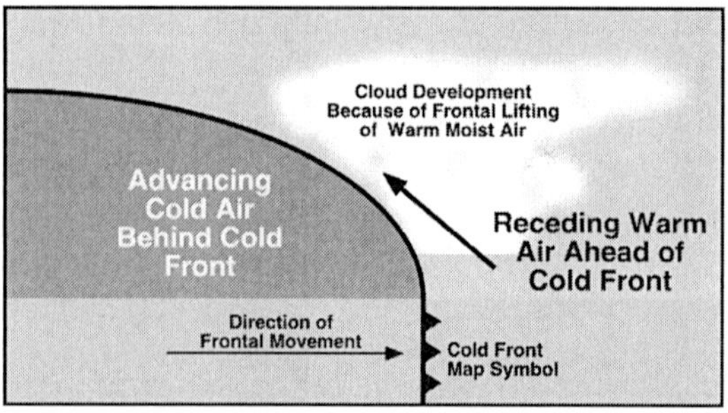

Fig. Atmospheric cross-section of a cold front.

A warm front is illustrated in the cross-section diagram below. A warm front is the transition zone in the atmosphere where an advancing warm subtropical, moist air mass replaces a retreating cold, dry polar air mass. On a weather map, a warm front is drawn as a solid red line with half-circles. The position of the half-circles shows the direction of frontal movement. Warm fronts move about 10 kilometers per hour in a northeast direction. This is less than half the speed of a cold front. The formation of clouds and precipitation ahead of the frontal zone is caused by gradual frontal lifting. High altitude cirrus, cirrostratus and middle altitude altostratus clouds are found well in advance of the front. About 600 kilometers ahead of the front, nimbostratus clouds occur. These clouds produce precipitation in the form of snow or rain. Between the nimbostratus clouds and the surface location of the warm front, low altitude stratus clouds are found. Finally, a few hundred kilometers behind the front scattered stratocumulus are common in the lower troposphere. Table 7r-2 describes some of the weather conditions associated with a warm front.

Table . Weather conditions associated with a warm front.

Weather Phenomenon	Prior to the Passing of the Front	Contact with the Front	After the Passing of the Front
Temperature	Cool	Warming suddenly	Warmer then leveling off
Atmospheric Pressure	Decreasing steadily	Leveling off	Slight rise followed by a decrease
Winds	South to southeast	Variable	South to southwest
Precipitation	Showers, snow, sleet or drizzle	Light drizzle	None
Clouds	Cirrus, cirrostratus, altostratus, nimbostratus, and then stratus	Stratus, sometimes cumulonimbus	Clearing with scattered stratus, sometimes scattered cumulonimbus

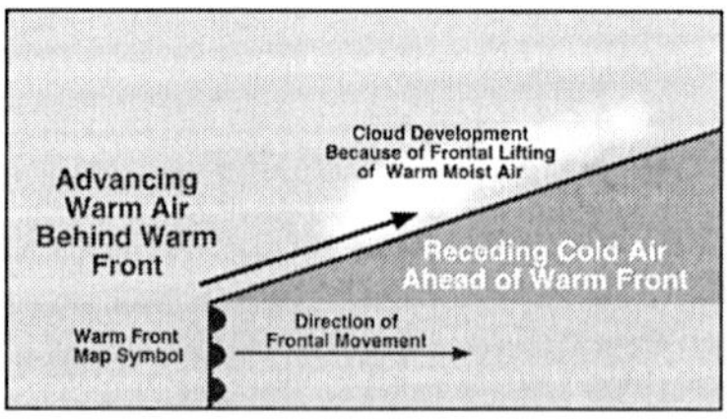

Fig. Atmospheric cross-section of a warm front.

Occluded fronts are produced when a fast moving cold front catches and overtakes a slower moving warm front. Two types of occluded fronts are generally recognized. A cold type occluded front occurs when the air behind the front is colder than the air ahead of the front. When the air behind the front is warmer than the air ahead of the front a warm type occluded front is produced. Warm type occlusions are common on the west coast of continents and generally form when maritime polar air collides with continental polar or arctic air. The cross-section diagram in Figure 7r-4 illustrates a cold type occlusion. Note that in the occlusion process the invading mild moist air that was found behind the warm front has been lifted into the upper troposphere.

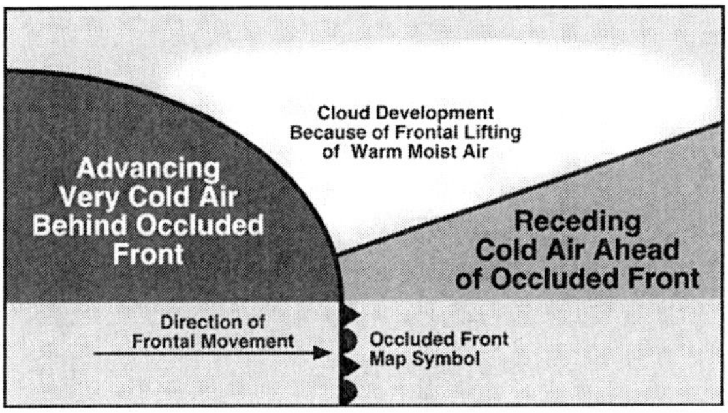

Fig. Atmospheric cross-section of a occluded front.

Finally, the frontal systems described on this page are often associated with a storm system known as the mid-latitude cyclone.

2

Forestry : An Introduction

Although most people know what a forest is, a definition of it which suits all cases is by no means easy to give. Manwood, in his treatise of the *Lawes of the Forest*, defines a forest as "a certain territory of woody grounds, fruitful pastures, privileged for wild beasts and fowls of forest, chase and warren, to rest and abide in, in the safe protection of the king, for his princely delight and pleasure."

This primitive definition has, in modern times, when the economic aspect of forests came more into the foreground, given place to others, so that forest may, in a general way, now be described as " an area which is for the most part set aside for the production of timber and other forest produce, or which is expected to exercise certain climatic effects, or to protect the locality against injurious influences." As far as conclusions can now be drawn, it is probable that the greater part of the dry land of the earth was, at some time, covered with forest, which consisted of a variety of trees and shrubs grouped according to climate, soil and configuration of the several localities. The conditions for an uninterrupted regeneration of the forest were favourable and the result was vigorous production by the creative powers of soil and climate.

Then came man and by degrees interfered, until in most countries of the earth the area under forest has been considerably reduced. The first decided interference was probably due to the establishment of domestic animals; men burnt the forest to obtain pasture for their flocks. Subsequently similar measures on an ever-increasing scale were employed to prepare the land for agricultural purposes. More recently enormous areas of forests were destroyed by reckless cutting and subsequent firing in the extraction of timber for economic purposes.

It will readily be understood that the distribution and character of the now remaining forests must differ enormously. Large portions of the earth are still covered with dense masses of tall trees, while others contain low scrub or grass land, or are desert. As a general rule, natural forests consist of a number of different species intermixed; but in some cases certain species, called gregarious, have succeeded in obtaining the upper hand, thus forming more or less pure forests of one species only. The number of species differs very much.

In many tropical forests hundreds of species may be found on a comparatively small area, in other cases the number is limited. Burma has several thousand species of trees and shrubs, Sind has only ten species of trees.

Central Europe has about forty species and the greater part of northern Russia, Sweden and Norway contains forests consisting of about half a dozen species. Elevation above the sea acts similarly to rising latitude, but the effect is much more rapidly produced. Generally speaking, it may be said that the Tropics and adjoining parts of the earth, wherever the climate is not modified by considerable elevation, contain broad-leaved species, palms, bamboos, andc. Here most of the best and hardest timbers are found, such as teak, mahogany and ebony.

The northern countries are rich in conifers. Taking a section from Central Africa to North Europe, it will be found that south and north of the equator there is a large belt of dense hardwood forest; then comes the Sahara, then the coast of the Mediterranean with forests of cork oak; then Italy with oak, olive, chestnut, gradually giving place to ash, sycamore, beech, birch and certain species of pine; in Switzerland and Germany silver fir and spruce gain ground. Silver fir disappears in central Germany and the countries around the Baltic contain forests consisting chiefly of Scotch pine, spruce and birch, to which, in Siberia, larch must be added, while the lower parts of the ground are stocked with hornbeam, willow, alder and poplar.

In North America the distribution is as follows: Tropical vegetation is found in south Florida, while in north Florida it changes into a subtropical vegetation consisting of evergreen broad-leaved species with pines on sandy soils. On going north in the Atlantic region, the forest becomes temperate, containing deciduous broad-leaved trees and pines, until Canada is reached, where larches, spruces and firs occupy the ground. Around the great lakes on sandy soils the broad-leaved forest gives way to pines. On proceeding west from the Atlantic region the forest changes into a shrubby vegetation and this into the prairies.

Farther west, towards the Pacific coast, extensive forests are found consisting, according to latitude and elevation above the sea, of pines, larches, fir, Thujas and Tsugas. In Japan a tropical vegetation is found in the south, comprising palms, figs, ebony, mangrove and others. This is followed on proceeding north by subtropical forests containing evergreen oaks, *Podocarpus,* tree-ferns and, at higher elevations, *Cryptomeria* and *Chamaecyparis.* Then follow deciduous broad-leaved forests and finally firs, spruces and larches. In India the character of the forests is governed chiefly by rainfall and elevation.

Where the former is heavy evergreen forests of Guttiferae, Dipterocarpeae, Leguminosae, Euphorbias, figs, palms, ferns, bamboos and india-rubber trees are found. Under a less copious rainfall deciduous forests appear, containing teak and sal (*Shorea robusta*) and a great variety of other valuable trees. Under a still smaller rainfall the vegetation becomes sparse, containing acacias, *Dalbergia sissoo* and Tamarix. Where the rainfall is very light or *nil,* desert appears. In the Himalayas, subtropical to arctic conditions are found, the forests containing, according to elevation, pines, firs, deodars, oaks, chestnuts, magnolias, laurels, rhododendrons and bamboos. Australia, again, has its own particular flora of eucalypts, of which some two hundred species have been distinguished, as well as wattles. Some of the eucalypts attain an enormous height.

Utility of Forests

In the economy of man and of nature forests are of direct and indirect value, the former chiefly through the produce which they yield and the latter through the influence which they exercise upon climate, the regulation of moisture, the stability of the soil, the healthiness and beauty of a country and allied subjects.

The *indirect* utility will be dealt with first. A piece of land bare of vegetation is, throughout the year, exposed to the full effect of sun and air currents and the climatic conditions which are produced by these agencies. If, on the other hand, a piece of land is covered with a growth of plants and especially with a dense crop of forest vegetation, it enjoys the benefit of certain agencies which modify the effect of sun and wind on the soil and the adjoining layers of air. These modifying agencies are as follows:

- The crowns of the trees intercept the rays of the sun and the falling rain; they obstruct the movement of air currents and reduce radiation at night.
- The leaves, flowers and fruits, augmented by certain plants which grow in the shade of the trees, form a layer of mould, or humus, which protects the soil against rapid changes of temperature and greatly influences the movement of water in it.
- The roots of the trees penetrate into the soil in all directions and bind it together. The effects of these agencies have been observed from ancient times and widely differing views have been taken of them.

Of late years, however, more careful observations have been made at so-called parallel stations, that is to say, one station in the middle of a forest and another outside at some distance from its edge, but otherwise exposed to the same general conditions. In this way, the following results have been obtained:

- Forests reduce the temperature of the air and soil to a moderate extent and render the climate more equable.
- They increase the relative humidity of the air and reduce evaporation.
- They tend to increase the precipitation of moisture. As regards the actual rainfall, their effect in low lands is *nil* or very small; in hilly countries it is probably greater, but definite results have not yet been obtained owing to the difficulty of separating the effect of forests from that of other factors.

- They help to regulate the water supply, produce a more sustained feeding of springs, tend to reduce violent floods and render the flow of water in rivers more continuous.
- They assist in preventing denudation, erosion, landslips, avalanches, the silting up of rivers" and low lands and the formation of sand dunes.
- They reduce the velocity of air-currents, protect adjoining fields against cold or dry winds and afford shelter to cattle, game and useful birds.
- They may, under certain conditions, improve the healthiness of a country and help in its defence.
- They increase the beauty of a country and produce a healthy aesthetic influence upon the people.

The *direct* utility of forests is chiefly due to their produce, the capital which they represent and the work which they provide. The principal produce of forests consists of timber and firewood. Both are necessaries for the daily life of the people. Apart from a limited number of broad-leaved species, the conifers have become the most important timber trees in the economy of man. They are found in greatest quantities in the countries around the Baltic and in North America. In modern times iron and other materials have, to a considerable extent, replaced timber, while coal, lignite and peat compete with firewood; nevertheless wood is still indispensable and likely to remain so.

This is borne out by the statistics of the most civilized nations. Whereas the population of Great Britain and Ireland, during the period 1880-1900, increased by about 20%, the imports of timber, during the same period, increased by 45%; in other words, every head of population in 1900 used more timber than twenty years earlier. Germany produced in 1880 about as much timber as she required; in 1899 she imported 4,600,000 tons, valued at £14,000,000 and her imports are rapidly increasing, although the yield capacity of her own forests is much higher now than it was formerly. Wood is now used for many purposes which formerly were not thought of. The manufacture of the wood pulp annually imported into Britain consumes at least 2,000,000 tons of timber.

A fabric closely resembling silk is now made of spruce wood. The variety of other, or minor, produce yielded by forests is very great and much of it is essential for the well-being of the people and for various industries. The yield of fodder is of the utmost importance in countries subject to periodic droughts; in many places field crops could not be grown successfully without the leaf-mould and brushwood taken from the forests. As regards industries, attention need only be drawn to such articles as commercial fibre, tanning materials, dye-stuffs, lac, turpentine, resin, rubber, guttapercha, andc. Great Britain and Ireland alone import every year such materials to the value of £12,000,000, half of this being represented by rubber. The *capital* employed in forests consists chiefly of the value of the soil and growing stock of timber. The latter is, ordinarily, of much greater value than the former wherever a sustained annual yield of timber is expected from a forest. In the case of a Scotch pine forest, for instance, the value of the growing stock is, under the above-mentioned condition, from three to five times that of the soil.

The rate of interest yielded by capital invested in forests differs, of course, considerably according to circumstances, but on the whole it may, under proper management, be placed equal to that yielded by agricultural land; it is lower than the agricultural rate on the better classes of land, but higher on the inferior classes. Hence the latter are specially indicated for the forest industry and the former for the production of agricultural crops. Forests require *labour* in a great variety of ways, such as:

- General administration, formation, tending and harvesting;
- Transport of produce;
- Industries which depend on forests for their prime material. The labour indicated under the first head differs considerably according to circumstances, but its amount is smaller than that required if the land is used for agriculture. Hence forests provide additional labour only if they are established on surplus lands. Owing to the bulky nature of forest produce its transport forms a business of considerable magnitude, the amount of labour being perhaps equal to half that employed under the first head. The greatest amount of labour is, however, required in the working up of the raw material yielded by forests. In this respect attention may be drawn to the chair industry in and around High Wycombe in Buckinghamshire, where more than 20,000 workmen are employed in converting the beech, grown on the adjoining chalk hills, into chairs and tools of many patterns.
  Complete statistics for Great Britain are not available under this head, but it may be mentioned that in Germany the people employed in the forests amount to 2.3% of the total population; those employed on transport of forest produce. 1%; labourers employed on the various wood industries, 8.6%; or a total of 12%. An important feature of the work connected with forests and their produce is that a great part of it can be made to fit in with the requirements of agriculture; that is to say, it can be done at seasons when field crops do not require attention. Thus the rural labourers or small farmers can earn some money at times when they have nothing else to do and when they would probably sit idle if no forest work were obtainable.

Whether, or how far, the utility of forests is brought out in a particular country depends on its special conditions, such as:

- The position of a country, its communications and the control which it exercises over other countries, such as colonies;

- The quantity and quality of substitutes for forest produce available in the country;
- The value of land and labour and the returns which land yields if used for other purposes;
- The density of population;
- The amount of capital available for investment;
- The climate and configuration, especially the geographical position, whether inland or on the border of the sea, andc. No general rule can be laid down, showing whether forests are required in a country, or, if so, to what extent; that question must be answered according to the special circumstances of each case.

These data exhibit considerable differences, since the percentage of the forest area varies from 3.5 to 50 and the area per head of population from 07 to 9.5 acres. Russia, Sweden and Norway may as yet have more forest than they require for their own population.

On the other hand, Great Britain and Ireland, Germany, Denmark, Portugal, Holland and even Belgium, France and Italy have not a sufficient forest area to meet their own requirements; at the same time, they are all sea-bound countries and importation is easy, while most of them are under the influence of moist sea winds, which reduces to a subordinate position the importance of forests for climatic reasons.

Intimately connected with the area of forests in a country is the state of ownership - whether they belong to the state, corporations or to private persons. Where, apart from the financial aspect and the supply of work, forests are not required for the sake of their indirect effects and where importation from other countries is easy and assured, the government of the country need not, as a rule, trouble itself to maintain or acquire forests. Where the reverse conditions exist and especially where the cost of transport over long distances becomes prohibitive, a wise administration will take measures to assure the maintenance of a suitable proportion of the country under forest. This can be done either by maintaining or constituting a suitable area of state forests, or by exercising a certain amount of control over corporation and even private forests. Such measures are more called for in continental countries than in those which are sea-bound, as is proved by the above statistics.

THE DEVELOPMENT OF FORESTRY

A feature of fundamental importance, which distinguishes forests from some other resources, is that they are renewable. This means that the mature forest, as it is cut, may be replaced by a new crop of trees; thus a continuous yield of wood is provided from the same area of land. The application of this principle is the primary basis of forestry practice, which may be defined briefly as the growing and using of successive crops of timber.

As the practice of forestry develops, the forests often come to be looked upon less as a natural resource and more as a crop that is dependent in large measure on man's knowledge and skill. However, forestry is not concerned exclusively with growing timber as a crop; forests contribute in many other ways to the well-being of mankind and proper forestry practices assure the maintenance and enlargement of these benefits. Man is not a conservationist by nature: history shows that as long as the existing forest appears to be adequate to the demand, it is subject to exploitation for immediate revenue and there is little concern for its future productivity. In other words, forest conservation, involving some investment of time and money, only begins when it is apparent that current practices are endangering the resource with critical effects upon the supply of forest products and services.

Down through the centuries, as man has increased in numbers and spread over the face of the earth, he has caused and is still causing widespread destruction of the forest cover, partly of necessity to provide land for agriculture, settlement and industrial development, partly through ignorance and greed and partly in the belief that the forest resource is inexhaustible.

This latter viewpoint is to some extent justified in the early development of countries and regions that have extensive forest wealth and relatively small populations. Here the forest is often looked upon as a hindrance to settlement and agricultural development, with the result that areas cleared may exceed requirements and include lands quite unsuitable for the purpose in mind.

Man has in fact retained under intensive agriculture only a small portion of the area that was originally cleared for this purpose. It is estimated that, as a result of his activities, man has destroyed about two thirds of the world's original forest cover; the remaining forests containing merchantable timber today occupy less than one fifth of the world's land area.

In some parts of the world forest destruction has been unchecked, despite the adverse effect that this has had on the general economy and on the living standards of the people. Asia, in particular, has suffered from the effects of forest clearance and forest destruction, with widespread soil deterioration, inadequate supplies of wood to meet basic domestic needs and low living standards for a large portion of the population. Elsewhere, as the danger became apparent, methods of managing the forests to ensure their continued productiveness were evolved and thus the science of forestry came into being. However, the philosophy of conservation has developed slowly and forestry has invariably been the child of necessity: only after periods of forest destruction and land clearance has the need for husbandry and planned forest economy been realized.

TRANSITION FROM EXPLOITATION TO MANAGEMENT

The transition from exploitation to management takes time and requires the solution of many problems, technical, economic and administrative. In the beginning it may be characterized by a strong reaction against spoliation aimed at preserving the forest intact — "locking it up" — for posterity. In general, however, this is the antithesis of sound management of a renewable resource, which for its fullest development requires periodic harvesting and regeneration; an exception would be the preservation of representative samples of natural forest for scientific study and as "wilderness" areas. Modern forestry originated in Germany and France and much of its development during the past two centuries has taken place in northwestern Europe. Attention to forest conservation in these countries came early owing to an increasing demand for timber and firewood and the difficulty of securing wood cheaply from other countries.

These conditions and the serious effects of deforestation on stream flow, particularly in montane regions, made evident an urgent need for forest management. During the nineteenth century expanding populations and the iron and coal economy of the industrial revolution created new demands for wood and gave added impetus to the development of forestry.

Slowly and usually only as the need became urgent, other countries have followed the lead of western Europe and developed forestry practices to meet their own requirements. Such developments usually begin with restriction on the use of the forest, as in the setting aside of timber reserves and with protection of the forest against its more serious enemies, particularly fire. These steps are followed by efforts to secure new growth after cutting, by encouraging natural reproduction and by planting: with this the practice of silviculture begins. Finally, the accessible, productive forests are managed under cutting regulations and suitable silvicultural techniques to ensure a continuous yield of timber in perpetuity. Such forests are said to be managed on a "sustained-yield" basis.

While the natural forest usually provides the starting point for a forestry programme, it may not represent the optimum conditions that an area is capable of and its perpetuation is not always looked upon as the ideal objective of forest management. The aim, rather, may be to improve the natural forest as far as possible within the limits set by the local environment, silvicultural knowledge and economic considerations and in view of the purpose for which the forest is to be used.

Attempts to go beyond these limits and completely change the natural forest pattern have sometimes failed. A well-known instance was the century-long attempt to convert the lowland hardwood forests of Saxony to pure Norway spruce.

Despite early rapid growth, the spruce was often badly damaged by snow and frost as well as by fungal and insect attacks. By the end of the third crop of spruce, conditions had so deteriorated that hardwood species had to be reintroduced and it was generally conceded that the plan had been unsuccessful.

On the other hand, there are many examples of the successful introduction of types and species of trees different from those in the natural forest. The success of these operations depends largely on careful study of the ecological factors involved.

The aim of forestry may vary from one country to another and even between regions within the same country, depending upon local conditions. Usually the primary purpose is to grow sufficient wood to meet a country's requirements — first for domestic consumption, then for export to foreign markets. Products other than wood, usually referred to as minor forest products, may sometimes be an objective of management. These vary in importance and kind from one country to another, *e.g.*, turpentine, resin, oils, fruits, maple syrup, bark for tanning, etc. The southeastern United States and the Landes in Gascony, France, are two regions where the production of resin and turpentine from species of pine is a major industry carried on in association with timber production.

While the production of wood and other commodities is often primary, in its broader aspects forestry is concerned with the interaction between the forest and its environment, as, for example, in the effects of the forest on the stabilization of soil, on the maintenance and restoration of soil fertility, on the regulation of stream flow and on the provision of a suitable environment for valuable fish and wild life. The protective influences of the forest are many and far-reaching and under certain conditions may be more important than the supplying of wood.

This is particularly true in some parts of the tropics and subtropics, in crucial drainage areas and in regions of light, sandy soil unsuitable for cultivation, where the removal of the forest cover often results in near-desert conditions. Forestry may also aim at the full development of the recreational value of an area; but even where recreation or protection are the main objectives, controlled cutting of wood can usually be carried on advantageously, if the cutting operations do not interfere with the primary purpose of management.

The term "multiple use" is widely used in referring to the management of a forest area for two or more purposes, one of which may be dominant, but not exclusive.

Bases of Forestry

Forestry may be defined as the scientific management of forests for the continuous production of goods and services. While its practice involves many problems of an economic and administrative nature, the solution of which is essential for success, its proper development must in the first place be based on a sound knowledge of the forests and of the laws that govern their development and on an understanding of the characteristics of the individual species that make up the forests and of the interrelationship of the forests with their environment.

The forests of the world include a great many species of trees, not necessarily having any close botanical relationship, with varying adaptations to climate, soil and topography and more or less distinct reproductive characteristics, forms and rates of growth, sizes and ages at maturity, light requirements and abilities to resist damage by insects and other destructive agencies. It is with an intimate knowledge of these features of the trees and tree communities that sound forestry practice must begin.

Supply of Timber : Imports and Exports

The net imports and exports of European countries (average data, calculated from the returns of recent years). The only timber-exporting countries of Europe are Russia, Sweden, Norway, Austria-Hungary and Rumania; all the others either have only enough for their own consumption, or import timber. Great Britain and Ireland import now upwards of 10,000,000 tons a year, Germany about 4,600,000 tons and Belgium about 1,300,000 tons.

Holland, France, Portugal, Spain and Italy are all importing countries, as also are Asia Minor, Egypt and Algeria. The west coast of Africa exports hardwoods and imports coniferous timber. The Cape and Natal import considerable quantities of pine and fir wood. Australasia Net Imports and Exports of European Countries. These net imports are received from non-European countries. They consist chiefly of valuable hardwoods, like teak, mahogany, eucalypts and others. Exports hardwoods and some Kauri pine from New Zealand, but imports larger quantities of light pine and fir timber.

British India and Siam export teak and small quantities of fancy woods. The West Indies and South America export hardwoods and import pine and fir wood. The United States of America will not much longer be a genuine exporting country, since they import already almost as much timber from Canada as they export. Canada exports considerable quantities of timber. The Dominion has still a forest area of 1,250,000 sq. m., equal to 38% of the total area and giving 165 acres of forest for every inhabitant. Although only about one-third of the forest area can be called regular timber land, Canada possesses an enormous forest wealth, with which she might supply permanently nearly all other countries deficient in material, if the governing bodies in the several provinces would only determine to stop the present fearful waste caused by axe and fire and to introduce a regular system of management. As matters stand, the supplies of the most valuable timber of Canada, the white or Weymouth pine, are nearly exhausted, the great stores of spruce in the eastern provinces are being rapidly destroyed and the forests of Douglas fir in the western provinces have been attacked for export to the United States and to other countries.

Taking the remaining stocks of the whole earth together, it may be said that a sufficient quantity of hardwoods is available, but the only countries which are able to supply coniferous timber for export on a considerable scale are

Russia, Sweden, Norway, Austria and Canada. As these countries have practically to supply the rest of the world and as the management of their forests is far from satisfactory, the question of supplying light pine and fir timber, which forms the very staff of life of the wood industries, must become a very serious matter before many years have passed.

Unmistakable signs of the coming crisis are everywhere visible to all who wish to see and it is difficult to over-state the gravity of the problem, when it is remembered, for instance, that 87% of all the timber imported into Great Britain consists of light pine and fir and that most of the other importing countries are similarly situated. In some of these countries little or no room exists for the extension of woodland, but this statement does not apply to Great Britain and Ireland, which contain upwards of 12,000,000 acres of waste land and 12, 500,000 acres of mountain and heath land used for light grazing. Onefourth of that area, if put under forest, would produce all the timber now imported which can be grown in Britain, that is to say, about 95% of the total.

Forest Management

In early times there was practically no forest management. As long as the forests occupied considerable areas, their produce was looked upon as the free gift of nature, like air and water; men took it, used it and even destroyed it without let or hindrance. With the gradual increase of population and the consequent reduction of the forest area, proprietary ideas developed; people claimed the ownership of certain forests and proceeded to protect them against outsiders. Subsequently the law of the country was called in to help in protection, leading to the promulgation of special forest laws. By degrees it was found that mere protection was not sufficient and that steps must be taken to enforce a more judicious treatment, as well as to limit the removal of timber to what the forests were capable of producing permanently.

The teaching of natural science and of political economy was brought to bear upon the subject, so that now forestry has become a special science. This is recognized in many countries, amongst which Germany stands first, closely followed by France, Austria, Denmark and Belgium. Of non-European countries the palm belongs to British India and then follow Ceylon, the Malay States, the Cape of Good Hope and Japan.

The United States of America have also turned their attention to the subject. Most of the British colonies are, in this respect, as yet in a backward state and the matter has still to be fought out in Great Britain and Ireland, though many writers have urged the importance of the question upon the public and the government. There can be no doubt that all civilized countries must, sooner or later, adopt a rational and systematic treatment of their forests.For details as to the separate countries, see the articles under the country headings; in this article only some of the more important countries are dealt with, in so far as the history of their forestry is important. A few notes on Germany and France will be given, because in these countries forest management has been brought to highest perfection; Italy is mentioned, because she has allowed her forests to be destroyed; and a short description of forestry in the United Kingdom and in India follows.

The winters being long and severe, an abundant supply of fuel is almost as essential as a sufficient supply of food. This necessity has led, along with a passion for the chase, to the preservation of forests and to the establishment of an admirable system of forest cultivation, almost as carefully conducted as field tillage. The Black Forest stretches the whole length of the grand-duchy of Baden and part of the kingdom of Wurttemberg, from the Neckar to Basel and the Lake of Constance. The vegetation resembles that of the Vosges; forests of spruce, silver fir, Scotch pine and, mingled with birches, beech and oak, are the chief woods met with. Until comparatively recent times large quantities of timber derived from these forests were floated down the Rhine to Holland and also shipped to England.

Now the greater part of it is used locally for construction, or it is converted into paper pulp. In the grand-duchy of Hesse the Odenwald range of mountains, stretching between the Main and the Neckar, contains the chief supply of timber. In the province of Nassau there are the large wooded tracts of the Taunus mountain range and the Westerwald.In Rhenish Prussia valuable forests lie partly in the Eifel, on the borders of Belgium and on the mountains overhanging the Upper Moselle, but they do not furnish such stately trees as the Black Forest and the Odenwald. The Spessart, near Aschaffenburg in Bavaria, is one of the most extensive forests of middle Germany, containing large masses of fine oak and beech, with plantations of coniferous trees, such as spruce, Scotch pine and silver fir.

Bavaria possesses other fine forest tracts, such as the Baierischewald on the Bohemian frontier, the Kranzberg near Munich and the Frankenwald in the north of the kingdom. North Germany has extensive forests on the Harz and Thuringian Mountains, while in East Prussia large tracts of flat ground are covered with Scotch pine, spruce, oak and beech.

Every German state has its forest organization: In Prussia the department is presided over by the Oberland Forstmeister at Berlin, while each province, or part of a province, has an Oberforstmeister, under whom a number of OberfOrsters administrate the state and communal forests. These, again, are assisted by a lower class of officials called Forsters. The Oberf6rsters throughout Germany are educated at special schools of forestry, of which in 1909 the following nine existed: In Prussia: at Eberswalde and Munden.

The National Forest Policy

The forest policy of the United States may be said to have had its origin in

1799 in the enactment of a law which authorized the purchase of timber suitable for the use of the navy, or of land upon which such timber was growing. It is true that laws were in force under the early governments of Massachusetts, New Jersey and other colonies, providing for the care and protection of forest interests in various ways, but these laws were distinctly survivals of tendencies acquired in Europe and for the most part of little use.

It was not until the apparent approach of a dangerous shortage in certain timber supplies that the first real step in forest policy was taken by the United States. Successive laws passed from 1817 to 1831 strove to give larger effect to the original enactment, but without permanent influence towards the preservation of the live oak, which was the object in view. A long period of inaction followed these early measures. *The unshaded areas are treeless, except along the Streams* 1831 the solicitor of the treasury assumed a partial responsibility for the care and protection of the public timber lands and in 18 55 this duty was transferred to the commissioner of the general land office in the Department of the Interior.

The effect of these changes upon forest protection was unimportant. When, however, at the close of the Civil War railway building in the United States took on an unparalleled activity, the destruction of forests by fire and the axe increased in a corresponding ratio and public sentiment began to take alarm. Action by several of the states slightly preceded that of the Federal government, but in 1876 Congress, acting under the inspiration of a memorial from the American Association for the Advancement of Science, authorized the appointment of an officer under the commissioner of agriculture, to collect and distribute information upon forest matters.

As the railways advanced into the treeless interior, public interest in tree-planting became keen. In 1873 Congress passed and later amended and repealed the timber culture acts, which granted homesteads on the treeless public lands to settlers who planted one-fourth of their entries with trees. Though these measures were not successful in themselves they directed attention towards forestry. The act which repealed them in 1891 contained a clause which lies at the foundation of the present forest policy of the United States.

By it the president was authorized to set aside " any part of the public lands wholly or in part covered with timber or undergrowth, whether of commercial value or not, as public reservations and the President shall, by public proclamation, declare the establishment of such reservations and the limits thereof." Some eighteen million :acres had been proclaimed as reservations at the time when, in 1896, the National Academy of Sciences was asked by the secretary of the interior to make an investigation and report upon " the inauguration of a rational forest policy for the forest lands of the United States." Upon the recommendation of a commission named by the Academy, President Cleveland established more than twenty-one million acres of new reserves on the 22nd of February 1897. His action was widely misunderstood and attacked, but it awakened a public interest in forest questions without which the rapid progress of forestry in the United States since that time could never have been made.

Within a few months after the proclamation of the Cleveland reserves the present national forest policy took definite shape. Under this policy the national government holds and manages, in the common interest of all users of the forests or its products, such portions of the public lands as have been set aside by presidential proclamation in accordance with the act of 1891.

These lands are held against private acquisition under the Homestead Act (except as to agricultural lands as hereafter mentioned), the Timber and Stone Act and other laws under which the United States disposes of its unappropriated public domain, but not against private acquisition under the Mineral Land Laws. They are selected from lands believed to be more valuable for forest purposes than for agriculture and are managed with the purpose of securing from them the best and largest possible returns, present and future, whether in the form of water for irrigation or power, of timber, of forage for stock, or of any other beneficial product. The aggregate area of the reserves, or national forests, has been steadily increased until they now include nearly all the timber lands left of the public domain.

The general lines of this policy were in part laid down by the commission already mentioned, in its report submitted to the secretary of the interior, May 1, 1897 and by the act of June 4, 1897, which was largely shaped by the work of the commission. Until this act was passed the national forests had been in theory closed against any form of use; nor had the possibility of securing forest preservation by wise use received much thought from those who had favoured their creation. Such a state of affairs could not continue.

Before long public opinion would have forced the opening to use of the resources thus :arbitrarily locked up and in the absence of any administrative system providing for conservative use, the national forests would inevitably have been abolished and the whole policy of government forest holdings would have ceased.

The act of June 4, 1897 was therefore of the first importance. This act conferred upon the secretary of the interior general powers for the proper management of the national forests through the general land office of his department. It provided for the designation and sale of dead, mature and large timber; authorized the secretary to permit free use of timber in small quantities by settlers, miners and residents; empowered him to "make such rules and regulations and establish such service as will insure the objects of such reservations, namely, to regulate their occupancy and use and to preserve the forests thereon from destruction "; and made violation of the act or of such rules and regulations a misdemeanour. The statute limited the power to establish forest reservations to the purpose of improving and protecting the forest, securing favourable conditions of water flows and furnishing a continuous supply of timber for the use and necessities of citizens of the United States.

Lands found, upon due examination, to be more valuable for other purposes than for forest uses might be eliminated from any reservation and all mineral lands within the reservations were left open to private appropriation under the mineral laws. The rights of settlers and claimants were safeguarded and civil and criminal jurisdiction, except so far as the punishment of offences against the United States in the reservations was concerned, was reserved to the States.

While the administration of the national forests was entrusted to the general land office, the same act assigned the surveying and mapping of them to the United States Geological Survey, which has published descriptions and maps of some of the more important.

No attempt was made in the general land office to develop a technical forest service. There were, indeed, at the time of passage of the act, less than ten trained foresters in the United States, no means of training more and very little conception of what forestry actually meant. The purpose of the administration was therefore mainly protection against trespass and fire, particularly the latter.

Regulations were made giving effect to the provisions of the act of June 4, set forth above, but in the absence of technical knowledge as to what might safely be done, the tendency was rather to restrict than to extend the use of the forest. Meanwhile, however, there was rapidly developing in another branch of the government service an organization qualified for actual forest management.

One year after the passage of the act of June 4, 1897, the division of forestry in the Department of Agriculture ceased to be merely a bureau of information and became an active agency for introducing the actual practice of forestry among private owners and for conducting the investigations upon which a sound American forest practice could be based. The work awakened great interest among forest owners and exerted a powerful educational influence upon the country at large.

The division extended its work and became the Bureau of Forestry. It drew into its employment for a time nearly all the men who were preparing themselves in increasing numbers (at first abroad, then in the newly-founded schools in the United States) for the profession of forestry and was soon recognized as qualified to speak authoritatively on technical questions connected with the administration of the national forests.

This led to a request from the secretary of the interior for the advice of the bureau on such questions. Working plans were accordingly undertaken for a number of the forests. The general land office, however, was not ready to attempt active forest management. Though some timber was sold and the grazing of stock regulated to some extent, the main object of the land office administration continued to be protection against fire. Many of the regulations which it made could not be enforced. The disadvantages of dispersal of the Federal government forest work among three separate agencies grew more and more apparent, until, on the 1st of February 1905, control of the 63,000,000 acres of forest reserves which up to that time had been set aside was transferred from the general land office to the Bureau of Forestry. In recognition of its new duties the designation of the bureau became the Forest Service.

Other provisions of the act which affected the transfer were that forest supervisors and rangers should be selected, so far as possible, from qualified citizens of the state or territory in which each forest was situated and that all money received from the sale of any products or the use of any land or resources of the national forests should be covered into the treasury and constitute a special fund for their protection, administration, improvement and extension.

Five days later a statute gave forest officers the power to arrest trespassers; and on the 3rd of March the lieu land selection law was repealed. This law had opened the way for grave abuses through the exchange of worthless land by private owners within the forests for an equal area of valuable timber lands outside.

The law has been modified since by the change of the old name " Forest Reserves " to " National Forests." The act of June iz, 1906, opened to homestead entry lands within national forests found by examination to be chiefly valuable for agriculture. The administration and improvement of the national forests are now provided for directly by congressional appropriation. The power to create national forests conferred on the president by the act of March 1891 has been repealed for the states of Washington, Oregon, Idaho, Montana, Wyoming and Colorado, but for no others.

The Forest Service began in earnest the development of all the resources of the national forests. Mature timber was sold wherever there was a demand for it and the permanent welfare of the forests and protection of the streams permitted, but always so as to prevent waste, guard against fire, protect young growth and ensure reproduction. Regulations were adopted which allowed small sales to be made without formality or delay, secured for the government the full value of timber sold and eliminated unnecessary routine.

Carc was taken to safeguard the interests of the government and provide for the maintenance of good technical standards. The conduct of local business was entrusted to local officers. Large transactions with general policies were controlled from Washington, but with careful 2 = *District Boundaries and Numbers* provision for first-hand knowledge and close touch with the work in the field.

Business efficiency and the convenience of the public were carefully studied. In short, an organization was created capable of handling safely, speedily and satisfactorily the complex business of making useful a forest property of vast extent, scattered through sixteen different states of an aggregate area of over 1,50o,000 sq. m. and with a population of 9,000,000. The growth since the 1st of July 1897 of the area of the national forests, of the expenditures of the government for forestry and of the receipts from the national forests, is shown by the statement which follows. Though the act of June 4, 1897, became effective immediately upon its passage, the fiscal year 1899 was the first of actual administration, because the first for which Congress made the appropriation necessary to carry out the law.

Forest is ripe for the axe, the demand is strong and control by trained men makes it safe to cut more freely. The increase is marked both in small and in large sales, but a score of sales for less than $5000 are made against one for more. The total cut is still far below the annual increment of the forests. As the demand grows restrictions must increase in order to husband the present supply until the next crop matures.

- The stumpage price would seem on the face of the figures to have risen from about one dollar to more than three dollars per thousand board-feet. The receipts, however, for any one year are not exclusively for the timber cut in that year, since payments are made in advance. In the year 1907 the average price obtained was something less than $2.50 per thousand. It is therefore true that stumpage prices have risen greatly, although conditions new to the American lumbermen are im *Area of National Forests, Annual Expenditures of the Federal Government for Forestry and National Forest Administration and Receipts from National Forests, 1898-1909* Until 1906, the sole source of receipts was the sale of timber. In the fiscal year 1907, however, timber sales furnished less than half the receipts. The following statement concerning the timber sales of the fiscal years 1904-1907 will serve to bring out the change that followed the transfer of control to the forest service in the midst of the fiscal year 1905:
  - A large excess each year in the amount of timber sold over that cut and paid for;
  - Nine times as much timber sold at the end of the four-year period as at the beginning and three times as much cut;
  - A much higher price obtained per thousand board-feet at the end of the period than at the beginning. Each of these matters calls for comment. The sales are of stumpage only; the government does no logging on its own account.
- More timber is sold each year than is cut and paid for, because many of the sales extend over several years. With increasing sales the amount sold each year for future removal has exceeded the amount to be removed during that year under sales of earlier years. Large sales covering a term of years are made because the national forests

contain much overmature timber, which needs removal, but which is frequently too inaccessible to be saleable in small amounts. To prevent speculation the time allowed for cutting is never more than five years and cutting must begin at once and be continued steadily.

- The volume of sales has increased rapidly because much 1 The United States fiscal year ends June 30 and receives its designation from the calendar year in which it terminates. Thus, the fiscal year 1898 is the year July 1, 1897 - June 30, 1898.

Administration transferred to Bureau of Forestry, February I, 1905. Full utilization of all merchantable material, care of young growth in felling and logging and the piling of brush, to be subsequently burned by the forest officers if burning is necessary, are among these conditions.

Timber to be cut must first be marked by the forest officers. Sales of more than $100 in value are made only after public advertisement.

Only the simplest forms of silviculture have as yet been introduced. The vast area of the national forests, the comparatively sparse population of the West, the rough and broken character of the forests themselves and the newness of the problems which their management presents, make the general application of intensive methods for the present impracticable.

Natural reproduction is secured. The selection system is most used, often under the rough and ready method of an approximate diameter limit, with the reservation of seed trees where needed. The tendency, however, is strongly towards a more flexible and effective application of the selection principle, as a better trained field force is developed and as market conditions improve.

One conspicuous achievement was the reduction of loss by fires on the national forests. During the unusually dry season of 1905 there were only eight fires of any importance and the area burned over amounted only to about 16 of 1% of the total area. In 1906 about 12 of 1% was burned. This was accomplished by efficient patrol, co-operation of the public and by preventive measures, such as piling and burning the brush on cut-over areas.

Since the beginning of 1906 the largest source of income from the national forests was their use for grazing. Stock-raising is one of the most important industries of the West. Formerly cattle and sheep grazed freely on all parts of the public domain. In the early days of the national forests the wisdom of permitting any grazing at all upon them was sharply questioned.

Unrestricted grazing had led to friction between individuals, the deterioration of much of the range through overstocking and serious injury to the forests and stream flow. The forests of the West, however, are largely of open growth and contain many grassy parks, the results of old fires and many high mountain meadows.

Under proper regulations the grass and other forage plants which they produce in great quantity can be used without detriment to the forests themselves and with great benefit to the stock industry, which often can find summer pasturage nowhere else. Except in southern California grazing is now permitted on all national forests unless the watersheds furnish water for domestic use; but the time of entering and leaving, the number of head to be grazed by each applicant and the part of the range to be occupied are carefully prescribed. Planted areas and cut-over areas are closed to stock until the young growth is safe from harm and goats are allowed only in the brushland of the foothills.

The results of regulation, in addition to the protection of forest growth and streams, are the prevention of disputes, improved range, better stock, stable conditions in the stock industry and the best use of the range in the interest of progress and development. The first right to graze stock on the forests is given to residents, small owners and those who have used the range before. Thus the crowding out of the weaker by the stronger and of the settler by the roving outsider has been stopped. In 1906 the forest service began to impose a moderate charge for the use of the national forest range.

The following statement shows the amount of stock grazed on the national forests 1904-09 and the receipts for the grazing charge: - A work of enormous magnitude which has now begun is planting on the national forests. At present, with low stumpage prices and incomplete utilization of forest products, clear cutting with subsequent planting is not practicable.

There are, however, many million acres of denuded land within the national forests which require planting. Such planting is still confined chiefly to watersheds which supply cities and towns with water. The first planting was done in 1892, in California. Since then similar work has been done on city watersheds in Colorado, Utah, Idaho and New Mexico. Other plantations are in the Black Hills national forest, where large areas of cut-over and burned-over land are entirely without seed trees and in the sandhill region of Nebraska. Up to 1908 about 2,000,000 seedlings had been planted, on over 2000 acres - a small beginning, but the work was entirely new and presented many hard problems.

The nursery operations of the forest service are concentrated at seven stations, located in southern California, Nebraska, Colorado, New Mexico Utah and Idaho, where stock is raised for local planting and for shipment elsewhere. These nurseries are small. Their annual productive capacity is between 8,000,000 and seedlings. Each nursery is practically an experimental forest-planting station, at which a large variety of species are grown and various methods are tried.

The organization of the administrative work of the national forests is by single forests. On the 1st of January 1908 the total number of forests was 165 with a total area of 162,023,190 acres. In charge of each forest is a forest supervisor. Under the supervisors are forest rangers and forest guards, whose duties include patrol, marking timber and scaling logs, enforcing the regulations and conducting some of the minor business arising from the use of the forests. Guards are temporary employes; rangers are employed by the year. The supervisors report directly to and receive instructions from the central office at Washington.

In this office there are four branches - operation, grazing, silviculture and products - each of which directs that part of the work which belongs to it, dealing directly with the supervisor.

For inspection purposes, however, the forests are separated into six districts, in each of which is located a chief inspector with a corps of assistants. The inspectors are without administrative authority, but assist by their counsel the supervisors and through inspection reports keep the Washington office informed of the condition of all lines of administrative work in progress. Administrative officers alternate frequently between field and office duties.

The number of forest officers in the several grades on the 1st of January 1908 were: 6 chief inspectors, 26 inspectors, 106 forest supervisors, 41 deputy forest supervisors, 820 forest rangers and 283 forest guards. The total number of employes of the forest service on the same date, including the clerical force, was 2034.

Besides the administration of the national forests, the forest service conducts general investigations, carries on an extensive educational work and co-operates with private owners who contemplate forest management upon their own tracts. This last work is undertaken because of the need of bringing forestry into practice, the lack of trained foresters outside of the employ of the government and the lack of information as to how to apply forestry and what returns may be obtained. Co-operation takes the form of advice upon the ground and, on occasion, of the making of working plans.

The educational work of the service is performed chiefly through publications, the purpose of which is to spread very widely a knowledge of the importance of forestry to the nation and of the principles upon which its practice rests. The investigations which the service conducts extend from studies. of the natural distribution and classification of American forests. and of their varied silvicultural problems to statistics of lumber production and laboratory researches which bear upon the economical utilization of forest products. As examples of these researches may be mentioned tests of the strength of timber, studies of the preservative treatment of wood for various uses, wood-pulp investigations and studies in wood chemistry.

Forestry Among the States

Among the states forestry has hardly reached the stage of practical application on the ground. New York holds 1,500,00o acres of forest land. It has a commission to care for its forest preserve and to protect the forest land throughout the state from fire. The constitution of the state, however, prohibits the cutting of timber on state land and thus confines the work entirely to protection of the forest and to the planting of waste areas. Pennsylvania is at present showing the most efficient activity in working out a forest policy. It has state forests of 820,000 acres, a good fire law more and more satisfactorily enforced and eight nurseries for growing planting material. In 1905, 160,000 white pine seedlings were set out. It has also a school for forest rangers, to be employed on the state forests and it has just established a state professional school of forestry.

Twenty-six of the states have regularly appointed forest officers, six have carried on studies of forest conditions in co-operation with the forest service and there is scarcely one which is not actively interested in forestry. Laws, generally good, to prevent damage from forest fires, have been enacted by practically all the states, but their enforcement has unfortunately been lax.

Public sentiment, however, is making rapid progress. Among the best laws are those of Maine, New Hampshire, Minnesota, New York, Pennsylvania and Wisconsin. The New York law, for example, provides for the appointment of one or more firewardens in each town of the counties in which damage by fire is especially to be feared. In other counties supervisors of towns are *ex-officio* fire-wardens. A chief fire-warden has general supervision of their work.

The wardens, half of the cost of whose services is paid by the state, receive compensation only for the time actually employed in fighting fires. They may command the service of any citizen to assist them. Setting fire to woods or waste lands belonging to the state or to another, if such fire results in loss, is punishable by a fine not exceeding $250 or imprisonment not exceeding one year, or both and damages are provided for the person injured. Since fire is beyond question the most dangerous enemy of forests in the United States, the measures taken against it are of vital importance.

The amount of forest land held by the different states and by the territory of Hawaii: A *rea of State Forest Reservations, 1907. Forestry on Private Lands. The* practice of forestry among private owners is of old date. One of the earliest instances was that of Jared Eliot, who, in 1730, began the systematic cutting of timber land to supply charcoal for an iron furnace at Old Salisbury, Connecticut. The successful planting of waste lands with timber trees in Massachusetts dates from about ten years later. But such examples were comparatively rare until recent times.

At present the intelligent harvesting of timber with a view to successive crops, which is forestry, is much more common than is usually supposed. Among farmers it is especially frequent. It was begun among lumbermen by the late E. S. Coe, of Bangor, Maine, who made a practice of restricting the cut of spruce from his forests to trees ro, 12 or sometimes even 14 in. in diameter, with the result that much of his land yielded, during his life, a second crop as plentiful as the first.

Many owners of spruce lands have followed his example, but until very recently without improving upon it. Systematic forestry on a large scale among lumbermen was begun in the Adirondacks during the summer of 1898 on the lands of Dr W. S. Webb and Hon. W. C. Whitney, of a combined area of over ioo,000 acres, under the superintendence of the then Division of Forestry. In these forests spruce, maple, beech and birch predominate, but the spruce alone is at present of tht first commercial importance.

The treatment is a form of the selection system. Under it a second crop of equal yield would be ripe for the axe in thirtyfive years. Spruce and pine are the only trees cut. The work had been executed, at least up to the year 1902, with great satisfaction to the owners and the lumbering contractors, as well as to the decided benefit of the forest. The lumbering is regulated by the following rules and competent inspectors are employed to see that they are rightly carried out:

- No trees shall be cut which are not marked.
- All trees marked shall be cut.
- No trees shall be left lodged in the woods and none shall be overlooked by the skidders or haulers.
- All merchantable logs which are as large as 6 in. in diameter at the small end must be utilized.
- No stumps shall be cut more than 6 in. higher than the stump is wide.
- No spruce shall be used for bridges, corduroy, skids, slides, or for any purpose except building camps, dams or booms, unless it is absolutely necessary on account of lack of other timber.
- All merchantable spruce used for skidways must be cut into logs and hauled out.
- Contractors must not do any unnecessary damage to young growth in lumbering; and if any is done, they must discharge the men who did it.

These two instances of forestry have been most useful and effective among lumbermen and other owners of forest land in the north-east. Among those which have followed their example are the Berlin Mills Paper Company in northern New Hampshire,, the Cleveland Cliffs Iron Company in northern Michigan and the Delaware and Hudson Railroad Company in New York, all of which have employed professional foresters.

The most notable instance of forestry in the south is on the estate of George W. Vanderbilt at Biltmore, N.C. This was the first case of systematic forestry under regular working plans in the United States. It was begun in 1891 on about 4000 acres and has since been extended until it now covers about ioo,000 acres. A professional forester with a corps of trained rangers under him is in charge of the work. The Pennsylvania Railroad has recently employed a trained forester and several assistants and has undertaken systematic forestry on a large scale. The effect of the work of the forest service in assisting private owners is evidenced by the fact that down to the year 1908 670 wood lots and timber tracts had been examined by agents of the forest service, of which 250 were tracts over 400 acres in extent and planting plans had been made for 436 owners covering a total area of 80,000 acres. Expert advice is also given to wood lot owners upon application by many of the state foresters.

American Practice

The conditions under which forestry is practised in Europe and in America differ so widely that rules which are received as axiomatic in the one must often be rejected in the other. Among these conditions in America are the highly developed and specialized methods and machinery of lumbering, the greater facilities for transportation and consequent greater mobility of the lumber trade, the vast number of small holdings of forest land and the enormous supply of low-grade wood in the timbered regions.

High taxes on forest properties, cut-over as well as virgin, notably in the north-western pineries and the firmly established habits of lumbermen, are factors of great importance. From these and other considerations it follows that such generally accepted essentials of European methods of forestry as a sustained annual yield, a permanent force of forest labourers, a permanent road system and the like, are in most cases utterly inapplicable in the United States at the present day in private forestry. Methods of forest management, to find acceptance, must there conform as closely as possible to existing methods of lumbering. Rules of marked simplicity, the observance of which will yet secure the safety of the forest, must open the way for more refined methods in the future.

For the present a periodic or irregular yield, temporary means of transport, constantly changing crews and an almost total ignorance of the silvics of all but a few of the most important trees - all combine to enforce the simplest silvicultural treatment and the utmost concentration of purpose on the two main objects of forestry, which are the production of a net revenue and the perpetuation of the forest. Such concentration has been followed in practice by complete success.

The forests with which the American forester deals are rich in species,usually endowed with abundant powers of reproduction and, over a large part of their range, greatly dependent for their composition and general character upon the action of forest fires. Of the commercially valuable trees there may be said to be, in round numbers, a hundred out of a total forest flora of about 500 species, but many trees not yet of importance in the lumber trade will become so hereafter, as has already happened in many cases. The attention of the forester must usually be concentrated upon the growth and reproduction of a single species and never of more than a very few. Thus the silvicultural problems which must be solved in the practice of forestry in America are fortunately less complicated than the presence of so many kinds of trees in forests of such diverse types would naturally seem to indicate.

The forest fire problem is one of the most difficult with which the American forester has to deal. It is probable that forest fires have had more to do with the character and distribution of forests in America than any other factor except rainfall. With an annual range over thousands of square miles, in many portions of the United States they occur regularly year after year on the same ground. Trees whose thick bark or abundant seeding gives them peculiar powers of resistance, frequently owe their exclusive possessions of vast areas purely to the action of fire.

On the economic side fire is equally influential. The probability, or often the practical certainty, of fire after the first cut, commonly determines lumbermen to leave no merchantable tree standing. Forest fires are thus the most effective barriers to the introduction of forestry.

Excessive taxation of timber land is another of almost equal effect. Because of it lumbermen hasten to cut and afterwards often to abandon, lands which they cannot afford to hold. This evil, which only the progress of public sentiment can control, is especially prevalent in certain portions of the white pine belt.

Forest Associations

Public sentiment in favour of the protection of forests is now widespread and increasingly effective throughout the United States. As the general understanding of the objects and methods of forestry becomes clearer, the tendency, formerly very marked, to confound ornamental tree planting and botanical matters with forestry proper is rapidly growing less. At the same time, the number and activity of associations dealing with forest matters is increasing with notable rapidity. There are now about thirty such associations in the United States. One of these, the Society of American Foresters, is composed exclusively of professional foresters. The American Forestry Association is the oldest and largest.

It has been influential in preparing the ground work of popular interest in forestry and especially in advocating and securing the adoption of the federal forest reservation policy, the most important step yet taken by the national government. It publishes as its organ a monthly magazine called *Forestry and Irrigation.* The Pennsylvania Forestry Association has been instrumental in placing that state in the forefront of forest progress. Its organ is a bi-monthly publication called *Forest Leaves.*

Other states which have associations or societies of special influence in forest matters are California, Massachusetts, Minnesota, Colorado, New Hampshire, Georgia and Oregon. Arbor Day, instituted in Nebraska in 1872 as a day for shade-tree planting by farmers who had settled on the treeless prairies, has been taken up as a means of interesting school children in the planting of trees and has spread until it is now observed in every state and territory. It continues to serve an admirable purpose.

Lumbering

All the operations of the lumber trade in the United States are controlled and to no small degree determined, by the peculiar unit of measure which has been adopted. This unit, the boardfoot, is generally defined as a board one foot long, one foot wide and one inch thick, but in reality it is equivalent to 14 4 cub. in. of manufactured lumber in any form. To purchase logs by this measure one must first know about what each log will yield in one-inch boards.

For this purpose a scale or table is used, which gives the contents of logs of various diameters and lengths in board feet. Under such a standard the purchaser pays for nothing but the saleable lumber in each log, the inevitable waste in slabs and sawdust costing him nothing.

In addition to this amount, an immense quantity of wood is used each year for fuel, posts and other domestic purposes and the total annual consumption is not less than 20 billion cub. ft. The years 1890 to 1906 were marked by rapid changes in the rank of the important timber trees with reference to the amount of timber cut and a shifting of the important centres of production.

Among coniferous trees, white pine has yielded successively to yellow pine and Douglas fir, while the scene of greatest activity has shifted from the Northern forest to the. Southern and from there is rapidly shifting to the Pacific Coast. The total cut of coniferous lumber has increased steadily, but that of the hardwoods is falling off and in 1906 it was 15% less than in 1899, while inferior hardwoods are gradually assuming more and more importance and the scene of greatest activity has passed from the middle west to the south and the Appalachian region. *Conifers. The* coniferous supply of the country is derived from four forest regions:

- The Northern forest;
- The Southern forest;
- The Pacific Coast forest;
- The Rocky Mountain forest.

The Northern forest was long the chief source of the coniferous lumber production in the United States. The principal timber tree of this region is the white pine, usually known in Europe as the Weymouth pine. It has an average height when mature of i to ft., with a diameter a little less than 3 ft., but the virgin timber is approaching exhaustion. White pine was one of the first trees to be cut extensively in the United States and Maine, the pine tree state, was at first the centre of production. In 1851 the cut of white pine on the Penobscot river was 144 million ft., that of spruce 14 million and of hemlock 11 million.

Thirty years later the pine cut had sunk to 23 million, spruce had risen to 118 million and hemlock had passed pine by a million feet. Meanwhile, the centre of production had passed from the north woods to the Lake States and for many years this region was the scene of the most vigorous lumbering activity in the world. Woods waste includes tops, stumps, cull logs and butts, but does not include defective trees left or trees used for road purposes. Mill waste includes bark, kerf, slabs and edgings.

The spruce (*Picea rubens*) is used chiefly for lumber, but it is in large and increasing demand in the manufacture of paper pulp. For the latter purpose hemlock, poplar and several other woods are also employed, but on a smaller scale. The total consumption of wood for paper in the United States for 1906 was 3,660,000 cords, of which 2,500,000 was spruce. Of this, however, 720,000 cords were imported from Canada.

- The chief product of the Southern forest is the yellow pine. This is the collective term for the longleaf, shortleaf, loblolly and Cuban pines. Of these the longleaf pine, called pitch-pine in Europe, is the most important. Its timber is probably superior in strength and durability to that of any other member of the genus *Pinus* and in addition to its value as a timber tree it is the source of naval stores in the United States. The average size of the mature longleaf pine is 90 ft. in height and 20 in. in diameter. Shortleaf and loblolly are other important members of this group. Their wood very closely resembles that of the longleaf pine and is often difficult to distinguish from it. The trees are also of about the same size and height. Loblolly is, however, of more rapid growth. The total cut of yellow pine in 1906 was 11,661,000,000 board ft.; it has perhaps not yet reached its maximum, but is certainly near it.

  Another important coniferous tree of the Southern forest is the bald cypress, which grows in the swamps. The cut in 1906 was 839,000,000 board ft., a gain of 69% over 1899.
- But the great supply of coniferous timber is now on the Pacific Coast. The Douglas fir, also known as Douglas spruce, red fir and Oregon pine, is the foremost tree in Oregon and Washington and the redwood in California. When mature the Douglas fir averages 200 ft. in height and 4 ft. in diameter and the redwood 225 ft. in height and 8 ft. in diameter. Other important trees of the Pacific Coast are sugar pine, western red cedar, western larch, Sitka spruce, western hemlock and western yellow pine.

  Logging on the Pacific Coast is characterized by the use of powerful machinery and by extreme skill in handling enormous weights. This is especially true in California, where the logs of redwood and of the big tree are often more than f ft. in diameter. Logging is usually done by wire cables operated by donkey-engines. The journey to the mill is usually by rail. The mills are often of great size, built on piles over tide water and so arranged that their product is delivered directly from the saws and dry kilns to vessels moored alongside. The products of the Pacific Coast forest make their way over land to the markets of the central and eastern states and into foreign markets. Among the lumber-producing states, Washington has in seven years jumped from fifth place to first and its output has increased from 1,428,000,000 board ft. in 1899 to 4,305,000,000 ft. in 1906. Oregon and California have increased their output from 734,000,000 each in 1899 to 1,605,000,000 and 1,349,000,000 ft. respectively in 1906. Of the total output of these three states (7,259,000,000 ft.) 4,880,000,000 ft. is Douglas fir and 660,000,000 redwood.

- The important lumber trees of the Rocky Mountain forest are the western yellow pine, the lodgepole pine, the Douglas fir and the Engelmann spruce. The Douglas fir, here extremely variable in size and value, reaches in this region average dimensions of perhaps 80 ft. in height by 2 ft. in diameter, the western yellow pine 90 ft. by 3 ft. and the Engelmann spruce 60 ft. by 2 ft. Mining, railroad and domestic uses chiefly absorb the annual timber product, which is considerable in quantity and of vast importance to the local population.

RIGHTS OVER FOREST RESOURCES

The tribal people are facing serious problems with regard to utilisation and rights over forest resources. Due to the increasing pressure on forests by various interest groups, there is a corresponding pressure on the tribals to reduce their dependency on forests. This is creating serious situations of conflict, as tribal life is symbiotic with land and forests and their livelihood and culture are based on their relationship with the natural wealth around them. The tribals are being harassed for using forestlands and being evicted in many places.

Such reports have come from places like Khammam, Visakhapatnam, Vizianagaram, Adilabad, Srisailam and other places. In Khammam in one particular village, the forest and police departments allegedly branded the tribals on their shoulders as an indication that they were destroying the forests. There are ambiguities in forest-revenue land demarcation. In some places like Nellore district, a lot of land on Velugonda hills is indicated as 'poramboku' in revenue records and as RF as per the forest department. These lands do not have any forest growth. The FD is taking up palm oil plantation in these RF lands (which is not a forest species). However, landless tribals are booked in criminal cases or prohibited from using these lands for agricultural cultivation. These lands should be given to tribals with pattas for cultivation.

In some areas like in Visakhapatnam district, the lack of clear forest boundaries is making tribals vulnerable to the exploitation of both the forest and revenue departments. A joint survey and demarcation of boundaries by both departments should be immediately taken up to arrest these conflicts. Such Joint surveys need to involve the villagers at various levels. The Forest-Revenue Boundary dispute is a perpetual problem in Adilabad and Warangal districts, leading to booking of cases by the Forest department and tension in these tribal villages.

In Visakhapatnam Agency and Nallamala areas, there is the unique problem of Enclosure Villages. There were many tribal villages that were not enumerated in the forest surveys. Due to such sheer negligence, the villages were not given revenue status and to this day, they do not have pattas for their lands. They face constant harassment from local forest officials, as they do not possess land records. Recognizing these enclosure villages and issuing pattas to tribals should immediately resolve this problem. In Buttapur (Adilabad district) and in Nellore Dt. (Yanadis), the tribals were given lands decades ago under the social forestry scheme and are cultivating there. But due to lack of pattas, they are being harassed by the police and forest departments and also do not have access to bank loans as they cannot prove their ownership. These tribals have to be given pattas as promised.

Attacks on tribals, their properties and livestock by wildlife are not compensated by the Forest Department. Several cases are pending where tribals have been either killed or disabled and yet have not received any monetary compensation as due to them under the Wildlife Act. The tribals should not be prohibited from entering the forests to collect NTFP for their domestic requirements, like firewood, medicinal herbs, food, or agricultural and housing material.

The tribals should not be defined as 'encroachers', as is being projected by the forest department and also as indicated in the Circular of IG Forests, MoEF dated 3rd May 2002. In districts like Visakhapatnam, East Godavari and Vizianagaram, where there is high prevalence of podu cultivation, the tribals are facing threats of eviction from the forest department. There was a notification issued by the government of A.P in 1987 ordering that pre-1980 settlements will not be evicted until further orders. This should be implemented. The JFM programme now renamed as the Community Forest Management programme of the A.P Forest Department has caused grievous violations with regard to tribal rights. One major violation is the displacement of tribals from their podu lands by reclaiming them back into the forests through the JFM programme. The official reports of the forest department and the World Bank (which has funded the project) reveal that 37,000 hectares of forestland has been reclaimed back from the people§. The APFD further states its intention of displacing tribals in G.O. where it calls for voluntary surrender of lands by involving NGOs to motivate people under the CFM programme. The APFD has come up with a proposal for rehabilitation of tribals displaced from their podu lands through a monetary and schematic approach. This should be condemned as it bypasses the issue of eviction and rights of people over their podu lands. Their act of reclaiming lands under JFM programme and the APFD's forestry project under World Bank assistance has to be scrutinized closely by the Commission.

Samata and a few other NGOs raised this issue by writing to the World Bank on its violations of its own Operational Directives 4.2 and 4.3 in its financial support to the APFD. This has led to the Bank's insistence on an RandR policy to be proposed and implemented by the APFD in its CFM programme

In Srisailam area, the Rajiv Gandhi Tiger Sanctuary has led to eviction of tribals from their original homes. They have not been properly settled so far. The concept of EDC (Eco Development Committees) that was introduced is working to the detriment of the tribals, as the objective of this programme was to reduce the forest dependency of the tribals. The tribals are given income-generating programmes and alternate sources of firewood by the forest department. This is an artificial mechanism that is not sustainable in the long term.

In the Srisailam Tiger Sanctuary area, the Chenchus, who are traditional hunter-gatherers, go into the forest everyday for all their needs. They are being harassed by the forest department for trespassing into the sanctuary. The groups who came from the chenchu area felt that the Chenchus should atleast be given identity cards to prevent harassment from forest department and the police department who mistake them for naxalites.

Another serious problem with regard to Sanctuaries is the settlement of people's rights. The Settlement Officer is the Conservator of Forests whereas the forest department is one of the interested parties in the land acquisition for the sanctuary and hence it is not appropriate to make the forest officer the settlement officer. It should be the revenue department (District Collector) who should be authorised to settle the people's rights. G.O.No.112 which calls for involvement of private industries like ITC, Reliance and others for taking up commercial plantations through the VSS should be withdrawn immediately, as this is a backdoor method of allowing private industries to enter forest and tribal lands. Such tripartite agreements between the government, the industries and the tribals can never provide a level playing ground for the tribal people and make them more vulnerable to the exploitation of private industries.

The historical injustices to the tribal people in this region and that of the tribals in southern Orissa due to construction of several 'development' projects and industries like Nalco, HAL, Sileru, Machkund and others which have displaced tribals in large numbers without any rehabilitation, and has forced them to migrate in search of livelihood. They are now being treated as encroachers and criminals in forestlands. This definition should be condemned and Commission should recommend proper mechanisms for resolution of conflicts. The Commission should strongly recommend the orders passed by the Ministry of Environment and Forests in 1990, on the recommendations of the Commissioner of Scheduled Castes and Scheduled Tribes, to ensure that such conflicts are resolved. Infact, the first two orders alone show a way forward to settle disputes concerning forestlands.

All the groups consulted strongly felt that the pressure on forests and the destruction of forests is taking place more due to increase in industries and private commercial activities like Mining, paper-mills, timber smuggling and the growing non-tribal population settling down in Agency areas, than due to the podu cultivation practiced by tribals. Eco-Tourism is also causing destruction of forests. For example, the A.P Tourism Department built huge infrastructure close to Farhanbad, near Mannanur (Srisailam) for Tourism purposes in the middle of the protected area and in Araku, Borra and Anantagiri (Visakhapatnam). While tribals are prohibited from entering these forests for their survival needs, it is totally unjust to allow tourism, which is causing a lot of degradation. Hence, the constitutional provisions of the Fifth Schedule, the Samatha Judgement and the PESA Act should be strictly implemented following the right spirit with regard to industries and non-tribal settlers, which includes the Tourism Department.

The Tourism Department also conducts tours through tribal villages, especially in Chenchu settlements, and exhibits them like museum pieces. The groups felt that this degrading exercise further underscored the lack of empathy among the various government departments towards the tribals.

Because of a shift in the policy of the government, profits to G.C.C. are more important than benefits to tribals. Although there are 32 items in the procurement list of GCC Ltd, it procures a limited number of commodities and at very low prices compared to the market rates. The tribals face several constraints in collection of NTFP and marketing these items because of the monopoly rights enjoyed by GCC Ltd. They do not have the right to sell their produce to private traders even when their rates are higher. Hence, a long-standing demand has been for removal of the monopoly restriction of GCC Ltd and instead a minimum support price should be provided by GCC to protect the tribals from exploitation of traders. The monopoly of forest department over Beedi leaf collection and sale should be removed. The tribals or VSS should be given 100% rights over income from beedi leaf and also the right of sale, with the forest department providing the support price.

INDIAN FOREST LAW AND POLICY

After Independence, however, and according to the newly enacted Indian Constitution, forests were placed under the so-called 'state list' according to which state legislatures have a primary right to make laws. In 1976, the Indian Forest Act was added to the concurrent list of the Constitution of India, giving the Central Government and states shared responsibility and control over forest matters. Thereby the Government of India does have the power to legislate on forestry issues but only after consulting the states. The responsibility for administering the forests lies primarily with the state government. The relative position of the Government of India depends on several factors, such as its contribution to the state budget and state polities. The balance of power between central and state governments has remained a key issue in forest management ever since.

After 1947, in a post-independent India, commercial exploitation and degradation of India's forests increased dramatically. Indeed, the 1952 National Forest Policy set out guidelines which were, for the most part, directed towards the supply of cheap timber and non-timber forest products for state-sponsored industrialization and modernization. In 1980, with the passage of the Forest Conservation Act, the central government reasserted some of its control over forest-based resources.

The Act restricts the state government's power to de-reserve a forest, and it restricts the use of forestland for non-forestry purposes without the prior approval of the central government. Unilateral decisions from the centre have prevailed from then onwards. Only six centres were set up to monitor forests and conservation (as per the Act), which is obviously insufficient for effective regional implementation in a country the size of India. Thus the Forest Conservation Act of 1980 has been problematic for a number of reasons, and has achieved little improvement in the 'conservation' of India's forests.

The National Forest Policy of 1988 envisaged people's involvement in the development and protection of forests *for the first time*. It stipulated that the requirements of people living in and near forests for fuel wood, fodder and small timber should be treated as the top priority, and forest communities should be motivated to identify themselves with the development and protection of forests from which they derive benefits. A primary task of all agencies responsible for forest management, including the forest development corporations, should be to engage tribal peoples closely in the protection, regeneration and development of forests, as well as to provide gainful employment to people living in and around the forests.

However the 1988 National Forest Policy has never been translated into law. It remains essentially a broad statement of government intent and does little in the way of specifying any legal rights or duties owed to forest communities-especially the tribal/indigenous peoples.

NATIONAL FOREST POLICY OF JOINT FOREST MANAGEMENT

One outcome of the National Forest Policy was the launching of Joint Forest

Management (JFM) initiatives by all the States and Union Territories. However, there have been no accompanying changes in the legal framework for JFM, and in almost all states it has been implemented through administrative orders, which do not take the existing infrastructure into consideration. Forests continue to be looked upon as the property of the Government and the delegation of management responsibilities to local communities is merely an administrative arrangement, falling short of any real devolution of power to local communities. As an illustration. Prakash Kashwan has looked into JFM from various points of view and highlights some other problems. One of the most important findings is that JFM has been allowed only in degraded forest, but in order to bring more communities under the ambit of such programme, communities have had to actively pursue degradation of natural forest. Furthermore, JFM requires the demarcation of boundaries between villages, often for the first time, but poor implementation has resulted in many boundary conflicts.

Joint Forest Management (JFM) in the state of Jharkand;

- JFM in the state of Jharkand is based on the premise that people are to blame for the destruction of the forest. The objective, therefore, is gradually to diminish people's dependence on forests.
- The Resolution to introduce JFM has been made unilaterally by the Forest Department. There is no scope for consultation with the people, no effort at consensus building, and no mechanism for dissemination of information.
- The resources required for villagers' needs are to be extracted from the protected forest according to the micro/annual-working plan. There is no mention of the existing rights of the villagers to use the produce from these forests. This is a gross violation of customary rights.
- A 33% share in net profits was promised. Now the promise is to share 90% of net profits with the forest dwellers. The profits however, are from commercial productions, which are not based on local needs, thus indicating the focus on commercialization of forest. The JFM Resolution in Jharkhand is simply an extension of the undivided Bihar Forest Policy (Jharkhand used to be part of Bihar before it became a separate state), which adopted Joint Forest Management in 1990. Jharkhand Forest Department adopted the JFM Resolution on 27th September 2001
- The Forest Department claims to have formed 2034 Village Forest Management and Conservation Committees and Eco-Development Committees managing an area of 5,589 sq. km. in a remarkably short time span. However, local people emphatically rejected JFM. Apparently, the figure is manipulated and incorrect.

Eviction of communities as Forest Encroachers

Indigenous peoples feel that measures that were put in place in the name of protecting the forest have only deprived them of their resources. The JFM initiatives have failed from the point of view of indigenous peoples, and there has also been widespread eviction of communities from forests. The forest department uses any means available to alienate the forest communities from the forest, reduce their dependence on the forest in the name of forest protection and conservation, and invite private partners and multinational companies to negotiate directly with village committees to access forests for carbon trading, Clean Development Mechanisms and trade in eco-services. The direction of forest management is thus very far from anything that takes traditional forest related knowledge into account or that recognises the rights of indigenous peoples to maintain their traditional forest practices; rather, it aims to separate them from the forests, and sets up an externally imposed process of commodification and marketing of forest goods.

Indigenous Peoples' Rights and Forest Management

The Government of India is not a signatory to the ILO Convention 169 and as such does not recognize indigenous peoples. It does recognise "Scheduled Tribes", but not all indigenous groups are scheduled. Some of the indigenous groups are recognized by the Government of India as Scheduled Tribes and enjoy a degree of special status. Some of the areas where the Scheduled Tribes live have been demarcated as Scheduled areas and under the Indian Constitution Sixth Scheduled provisions, Autonomous District Councils have been set up for local administration of the Scheduled Tribes.

In theory there are some measures to protect the interests of Scheduled Tribes. The Constitution empowers the President of India and state governors to withhold any law considered detrimental to tribal/indigenous peoples' interests in Scheduled Areas. Items V and VI of the Indian Constitution give special privileges to the Scheduled Tribes and the Panchayats Act (Extension to the Scheduled Areas), 1996 (PESA) is also designed to offer some protection. However, in practice most of the laws restrict the rights of and control of forest communities. Specifically, the Indian Forest Act (IFA), Forest Conservation Act (FCA) and Wildlife Protection Act (WLPA) continue to be used to hound forest dwellers.

JFM and CFM are present in most Indian states, but are simply plantation programmes related to timber and other economic products, which have nothing in essence to do with forests. They should include Village Forest Committees directly representing the community living in or using the forest. These programmes are operational to varying degrees in different places with varying degrees of participation of indigenous peoples depending upon the local situation. The involvement of indigenous peoples is focused on providing them with a share in the economic benefits and *de facto* also acts to wean them away from their dependence on forests. In other words, integral to these programmes is a subtle (in some cases blatant) redefinition of indigenous peoples' relationship to the forests. The degree of incorporation of TFRK is varied, if there is any at all; the expertise is provided by the forest department, sometimes in partnership with NGOs.

Indigenous peoples' rights to the forests are progressively minimized. In some cases, indigenous peoples do receive some economic benefits (though in others non-indigenous peoples dominate the Village Protection Committees), and in the context of the takeover of resources necessary for basic survival and consequent deprivation, these economic benefits at times are seen as a steady source of survival income. This in certain cases has superficially reduced tensions and conflicts between the forest department and the indigenous peoples as a sense of co-existence is manufactured. However, this also means that the forest department is able to stifle protests and resistance. In many places, where the indigenous peoples' political consciousness is well developed, they keep away from or oppose these programmes leading to conflicts, as has happened in parts of Madhya Pradesh.

India's report to UNFF3 (2002) describes measures taken to make use of indigenous capacity and local knowledge in participatory forest management. Clearly, the picture painted by this report is not matched by the situation on the ground.

3

Forest: An Area of Density of Trees

The forest is a complex ecosystem consisting mainly of trees that buffer the earth and support a myriad of life forms. The trees help create a special environment which, in turn, affects the kinds of animals and plants that can exist in the forest. Trees are an important component of the environment. They clean the air, cool it on hot days, conserve heat at night, and act as excellent sound absorbers.

Plants provide a protective canopy that lessens the impact of raindrops on the soil, thereby reducing soil erosion. The layer of leaves that fall around the tree prevents Run-off and allows the water to percolate into the soil. Roots help to hold the soil in place. Dead plants decompose to form humus, organic matter that holds the water and provides nutrients to the soil. Plants provide habitat to different types of organisms. Birds build their nests on the branches of trees, animals and birds live in the hollows, insects and other organisms live in various parts of the plant. They produce large quantities of oxygen and take in carbon dioxide. Transpiration from the forests affects the relative humidity and precipitation in a place.

The FAO (Food and Agriculture Organisation) has defined forest as land with tree crown cover (or equivalent stocking level) of more than 10% and area of more than 0.5 hectare. The trees should be able to reach a minimum height of 5 m at maturity in situ. Forests are further subdivided into plantations and natural forests. Natural forests are forests composed mainly of indigenous trees not deliberately planted. Plantations are forest stands established by planting or seeding, or both, in the process of afforestation or reforestation.

Forests can develop wherever the average temperature is greater then 10 °C in the warmest month and rainfall exceeds 200 mm annually. In any area having conditions above this range there exists a variety of tree species grouped into a number of forest types that are determined by the specific conditions of the environment there, including the climate, soil, geology, and biotic activity. Forests can be broadly classified into types such as the taiga (consisting of pines, spruce, etc.), the mixed temperate forests (with both coniferous and deciduous trees), the temperate forests, the sub tropical forests, the tropical forests, and the equatorial rainforests. The six major groups of forest in India are moist tropical, dry tropical, montane sub tropical, montane temperate, sub alpine, and alpine. These are subdivided into 16 major types of forests.

India has a long history of traditional conservation and forest management practices. Under British rule, forest management systems were set in place mainly to exploit forests. Non-etheless, there were some attempts to conserve forests and meet the needs of local communities. The Indian National Forest Policy of 1894 provided the impetus to conserve India's forests wealth with the prime objectives of maintaining environmental stability and meeting the basic needs of the fringe forests user-groups. Consequently, forests were classified into four broad categories, namely forests for preservation of environmental stability, forests for providing timber supplies, forests for minor forest produce, and pasture lands. While the first two categories were declared as reserve forests, the rest were designated as protected forests and managed in the interests of the local communities

Soon after independence, rapid development and progress saw large forest tracts fragmented by roads, canals, and townships. There was an increase in the exploitation of forest wealth. In 1950 the Government of India began the annual festival of tree planting called the Vanamahotsava. Gujarat was the first state to implement it. However, it was only in the 1970s that greater impetus was given to the conservation of India's forests and wildlife. India was one of the first countries in the world to have introduced a social forestry programme to introduce trees in non-forested areas along road sides, canals, and railway lines. A forest is an area with a high density of trees. There are many definitions of a forest, based on the various criteria. These plant communities presently cover approximately 9.4% of the Earth's surface (or 30% of total land area) and function as habitats for organisms, hydrologic flow modulators, and soil conservers, constituting one of the most important aspects of the Earth's biosphere. Historically, "forest" meant an uncultivated area legally set aside for hunting by feudal nobility, and these hunting forests were not necessarily wooded much if at all.

However, as hunting forests did often include considerable areas of woodland, the word forest eventually came to mean wooded land more generally. A woodland is ecologically distinct from a forest. The latitudes 10° north and south of the Equator are mostly covered in tropical rainforest and the latitudes between 53°N and 67°N with boreal forest.

CHANGES IN FOREST COVER AND CONDITION

An analysis of changes in the cover and condition of the world's forests requires a differentiation between:

1 the increase in forest cover (by afforestation or natural colonisation of trees on non-forest land) or decrease (by deforestation) of forest area; and

2 changes in forest condition, either positive (recovery of degraded stands, stand improvement treatments) or negative (decline, defoliation or dieback, effects of forest fires, degradation through unsustainable exploitation for wood, overgrazing, effects of pests and diseases).

Changes in Forest Cover

Between 1980 and 1995, the extent of the world's forests decreased by some 180 million ha, an area about the size of Indonesia or Mexico. This represents a global annual loss of 12 million ha, an area equivalent to the size of Greece or Bangladesh. During this 15-year period, developing countries lost nearly 200 million ha of natural forests, mostly through clearing for agriculture (shifting cultivation, other forms of subsistence agriculture, the establishment of cash crop plantations such as oil palm, and ranching). This was only very partially compensated for by the establishment of new forest plantations. Over the same period, forests in the developed world expanded slowly (by some 20 million ha) through afforestation and reforestation, including natural regrowth on land abandoned by agriculture.

Forest plantation and natural regeneration on abandoned agricultural land more than compensated for clearing of forests due to urbanisation and infrastructure development in most industrialized countries, outside the former USSR. While the gain of forest cover in developed countries was 20 million ha in the period 1980-95, 9 million ha of this increase occurred in the 5 years from 1990 to 1995.

The situation is quite different in the developing world, where deforestation exceeded net afforestation/reforestation, particularly in the tropical zone. As a whole, the annual rate of deforestation in the developing world between 1990 and 1995 was 0.7%, equivalent to 12.6 million ha, with the highest rate in tropical Asia-Oceania, closely correlated with population and income growth. The lowland forest formations were the most affected by deforestation, although the proportional loss during this period was greater in upland formations (1.1% compared with 0.8% in the lowland formations). There is some evidence that the rate of loss of forest cover in developing countries was slowing towards the end of the 15-year period 1980-95. The annual rate of forest loss in the period 1980-90 was 15.5 million ha compared with 13.7 million ha between 1990 and 1995.

Conversion of Forests to other Land Cover

The Forest Resource Assessment 1990 carried out an assessment of the relative importance of the various factors involved in deforestation at regional and global levels between 1980 and 1990 in the whole tropical belt. Among the most significant outputs of the study were the 'area transition matrices' of the type, which indicate transfers from one land cover class to another over the 3068 million ha of the tropical zone covered by the sample. For instance, the first row shows that 1275.9 million ha of the total area of 1368 million ha of closed forest in 1980 remained as closed forest up to 1990 and also shows the fate of the 92.1 million ha converted to other land cover classes (open forest; long fallow; fragmented forest; shrubs and short fallow; other, *i.e.* non-wooded, land cover; and forestry or woody forest and agricultural tree plantations). All these transfers involved changes of woody biomass attempts to capture by replacing each class along the y-axis (average biomass per hectare) and showing the importance of transfers along the x-axis.

Changes in Forest Condition

Factors affecting the health and vitality of all types of forests have attracted increasing attention in recent years. Outbreaks of fires have made headline news in developed and developing countries alike, while insect and disease attack and generalised declines in forest condition have caused more localized concern. Insects and disease and forest fires exist under 'natural' systems and conditions; what has caused concern has been their spread in artificial systems and in conditions considerably modified by humankind, resulting in direct economic loss.

Fire arising from natural causes (*e.g.* lightning) has been a major influence and is a driving evolutionary force in several forest ecosystems, such as the pine forests of Central America. Fires are important to the health and maintenance of these ecosystems and the total exclusion of fire from them can lead to the build-up of fuel and ultimately abnormally destructive fires that cause the loss of the ecosystem, as happened with the fire exclusion policy of the National Parks Service of the USA and the disastrous fires of 1988.

However, many fires arising from human activities, whether planned or unplanned, have damaged and destroyed large areas of forests and woodlands. Resource managers face a demanding public often with conflicting needs and incomplete information, leading to the development of new policies that take into account public attitudes towards fire management.

At the global level, an agreement was signed at a meeting in 1990 to initiate the International Decade for Natural Disaster Reduction and the target date of 2000 was set for signatory countries to effectively monitor and manage wildfires in their respective countries. In view of the lack of information on causes, extent and the number of fires at present, and the difficulties in collecting data, this target date is unlikely to be met.

Developed Countries

Total forest area is slowly increasing in the developed world, although the need for protection from fire, insects and disease has continued and some aspects of forest condition give cause for concern. Although the widespread decline and death of certain European forests due to air pollution predicted by many in the 1980s did not occur, deteriorating forest condition has remained a serious concern in Europe and North America. The main causes of the decline of forest condition in Europe have been low soil moisture availability due to drought, and high temperatures. These factors, combined with airborne pollution and the use of non-adapted seed sources in forest plantation development, have predisposed forests to attack by insects and disease and to decline. Forest damage due to air pollution is severe in some parts of central and eastern Europe in particular. Air pollution is also said to have a role in the decline observed in certain species in other industrialised countries, such as sugar maple *(Acer saccharum)* in eastern Canada, the high-elevation forests of red spruce *(Picea)* in eastern USA and *Cryptomeria japonica* stands in Japan.

The exclusion of fire may have caused the decline of some forest ecosystems, such as oak stands in central USA or natural eucalypt forests in Australia where fire is an integral part of the ecology of some plant communities and here it may be used to facilitate their natural regeneration.

Despite an overall increase in the number of fires (by some 40% in Europe, 8% in North America and 120% in the former USSR), wildfires over large areas of forest have become less frequent in the industrialised countries. Further, the average area burned by wildfire decreased in most regions between 1983-90 and 1991-94. There has been a slight reduction in total areas burned in the developed world overall as a result of improved prevention, detection and control systems, although there has been an increase of 15% in the former USSR. In 1990, the average area of forest and other wooded land affected annually by fires in Europe, North America and the former USSR together was about 4.26 million ha or 0.22% of their total area of forest and other wooded land. Pests and diseases remain constant threats, particularly to the semi-natural forests and plantations in the industrialised world. Trees are more likely to be attacked when under stress, whether from abnormal weather conditions such as drought or high temperatures, airborne pollution, uncontrolled movement of germplasm or lack of management. Recent outbreaks of pests and diseases that may have been related to stress include the decline of various species of oaks in many parts of Europe and in central Russia, the recurrent infestation of forests in Poland by the nun moth, the attack on beech by the beech scale in western and central Europe, birch and ash diebacks in north-eastern USA, the littleleaf disease of shortleaf pine in southern USA, and the 'x-disease' of pines in southern California.

FOREST CONDITIONS IN DEVELOPING COUNTRIES

Less attention has been paid to forest condition in developing countries because reduction in forest area has attracted more attention. This reduction in density has often been the result of over-exploitation for timber and fuelwood. Overgrazing and repeated bush fires are other significant causes of decrease in density, especially in the dry tropical and non-tropical zones.

Every year, very large areas of savannah woodland and mixed forest-grassland formations are affected by fires set by herders to provide an early flush of green grass when the rains start, particularly in the dry zones of Africa and South America, although no reliable data are available on their extent.

Forests in the humid tropics have also, at times, been affected by large fires, the most serious in recent years having been those associated with the dry weather conditions arising from the El Niño Southern Oscillation (ENSO). In 1997-98, these caused large-scale destruction of logged and secondary forest in Indonesia (particularly Kalimantan, Sumatra and Irian Jaya), Mexico and Central American countries, as well as leading to smoke pollution over wide areas beyond the borders of these countries. The cause of these fires has been burning for land clearance, but it should be noted that the fires are not necessarily in forest; although described as 'forest fires' they are frequently in grassland or on land that has been cleared of forest. It is *estimated* that of the 2 million ha burned in Indonesia in 1997, 150000-200 000 ha may have been in forest. Other serious fires associated with ENSO effects have included that in East Kalimantan, which burned 3.6 million ha in 1983.

Coniferous forests in the humid tropics have often been affected by fires, although it must be noted that fire is frequently necessary to maintain and regenerate them. In the 1980s, the area of pine forest in Honduras and Nicaragua burned annually amounted to some 65 000ha (or about 3.5% of the total pine forest area of these two countries), and widespread fires in natural and artificial tropical pine forests occur in other countries such as Guatemala, Mexico and Indonesia (northern Sumatra).

In the subtemperate and temperate zones of the developing world, fire is also a permanent threat to forests, particularly when they are no longer used by local people for grazing and other purposes. In the 1980s, the average area of forest and other wooded land burned annually was 140 000 ha in the temperate/ subtemperate zones of South America (including southern Brazil). From 1950 to 1990, fires in China are reported to have affected an average of 890 000 ha annually, the most damaging one having been the 'May 6' fire, which burned some 1.85 million ha in the north-eastern province of Heilongjiang in 1987. In the absence of a global statistical fire database, it is difficult to provide an overall estimate of the annual extent of fires in forests and other wooded lands.

A very crude estimate for the temperate/subtemperate and humid tropical zones of the developing world (leaving aside the significant dry tropical zone, for which little reliable information exists) would be of the order of 2 million ha of forest and other wooded land annually during the 1980s. Given the lack of sufficient capacity in fire prevention and control in most developing countries, no significant reduction in wooded areas burned is likely to occur in the near future.

Outbreaks of pests and diseases in developing countries are generally reported for those plantations and planted trees where the impact is most apparent. Introduced pests are usually extremely destructive and several of them have had very damaging effects in recent years in the developing world. Examples of introduced insects affecting plantations and planted trees include the *Leu-caena* psyllid, *Heteropsylla cubana,* which has spread into Asia and the Pacific islands and is now extending across Africa; the cypress aphid, *Cinara cupressi,* which is established in eight eastern and southern African countries and is causing heavy mortality among a number of exotic and indigenous species but especially the important plantation species *Cupressus lusitanica,* threatening its future role in the plantation programmes of these countries; and the European woodwasp, *Sirex noctilio,* which has spread into Argentina, Uruguay and southern Brazil, affecting principally *Pinus taeda* but which may become a threat for the large Chilean plantation estate of *Pinus radiata.*

Strategies to control introduced insects from attacking forest trees include regulatory measures, eradication, integrated pest management, and monitoring of population levels and occurrence of potentially harmful species.

Effective monitoring of insect populations will be especially important in view of the large plantation programmes being carried out to produce industrial roundwood, which often rely on one or a few exotic species. Such programmes may be able to afford the investment that is necessary for control measures. Smaller-scale programmes may have to insure against failure due to unexpected attack by insects or diseases by diversifying into several species adapted to the sites and end-uses.

There is also evidence of forest decline due to a combination of biotic and abiotic factors in the developing world, with air pollution likely to play an increasing role in some cases as industrial and transportation infrastructure develops. Examples of such decline include neem *(Azadirachta indica)* in the Sahel, framiré *(Terminalia ivoren-sis)* in Ivory Coast and Ghana, and *Eucalyptus globulus* plantations in Colombia and Peru, while airborne pollution is affecting forests near large cities and industrial areas in China (*e.g. Pinus massoniana* stands near Nanshan). Despite the overall significance of the corresponding damage and losses, surveys of forest decline and diebacks in developing countries remain all too rare.

RECENT ESTIMATES OF GLOBAL FOREST AREA

In 1995 forests were estimated to cover 3454 million ha, or 26.6% of the total land area of the world (Greenland and Antarctic excepted). Almost two-thirds of the world's forests were located in seven countries: Russia, Brazil,

Canada, USA, China, Indonesia and Zaire; 29 countries had more than half of their land covered by forest, of which 21 were in the tropical belt. However 49 countries, in addition to the many non-forested small island states and territories, had less than 10% of their land covered by forests. Five entire subregions were in this category: North Africa (1.2% of the land area), Near East (1.9%), Temperate Oceania (6.2%), Non-tropical Southern Africa (6.8%) and West Sahelian Africa (7.5%).

The latest information on the distribution of forest and of forest cover change by ecological zones was provided in 1990; the results of the Forest Resources Assessment 2000 were released by FAO in 2000. In 1990, temperate and boreal forests occupied 1.64 billion ha and tropical forests 1.76 billion ha. A breakdown by ecological zone was available only for tropical forests.The figures on tropical forests show that most (88%) are in lowlands; of these, tropical rain forests accounted for 47% of all lowland tropical forests, followed by moist deciduous forest (38%) and dry and very dry formations (15%).

Natural Forests

The forest area estimates described above include undisturbed forests, forests modified by humans through use and management (or 'seminatural' forests) and forests created artificially by humankind (*i.e.* forest plantations) by afforestation or reforestation. (Afforestation is defined as the establishment of a tree crop on an area from which it has always, or for a very long time, been absent. Reforestation is defined as the establishment of a tree crop on forest land.) In most industrialised countries, particularly in continental Europe, forests are being managed in such a way that at management-unit level a continuum exists from low-intensity management, involving natural regeneration, through more intensive methods involving some artificial planting to highly intensive methods with complete planting and cultivation; this makes it difficult to isolate figures for natural forest and plantations. The distinction between natural or seminatural forests and forest plantations can more easily be made for developing countries and some industrialised countries such as New Zealand in which forest plantations have been established using introduced species.

Interest in natural forests, particularly their role in the conservation of biological diversity, has led to efforts to compare forests today with what is thought to be their original character and to give complete protection to areas of forests which have had no, or minimal, human interference. Although there are difficulties in identifying the extent of natural forest, compounded by problems of definition, some information exists that can be used as an indication of broad patterns of natural forests in various regions.

An attempt has been made by the World Wide Fund for Nature (WWF) to quantify the area of forests in western Europe that has been relatively undisturbed by humans or which has retained much of its natural character.

Their report distinguishes between 'virgin forest', defined as 'forest ecosystems whose characteristics are determined exclusively by natural location and environmental factors... without human influences present or visible any more', and 'natural and ancient seminatural forests', which 'have not been planted or sown by man for the past two centuries' and 'which continue to have a large number of the natural elements'. The study found that only a small proportion (probably < 1%) of the total forest land in northern and western Europe could be considered as virgin forest, which has arisen since the last glaciation. Almost all was located in Sweden, Finland and Norway, with small areas in Greece, Austria and Switzerland and (according to another author) in France. In eastern Europe, Slovakia and Belarus have considerable areas of virgin forest while Poland and Croatia have small areas. In addition, in northern and western Europe, the WWF report identifies natural and ancient seminatural forests representing 2.1% of the total forest cover (1990) of the 16 countries concerned. There are a further 3 million ha of land in the region in national parks and other protected areas and another 50 000 ha in small forest reserves (mostly for nature conservation and scientific research), whose use is tightly restricted.

The situation in temperate and boreal North America is quite different from that of densely populated Europe and Japan, where use and management of forests for many centuries have left very little of the original forest area untouched. 'Old growth forests', as they are called in North America, still cover extensive areas.

On the lands managed by the US National Park Service alone, old-growth forests covered 1.97 million ha in 1988. Although not a direct measure of the extent of old growth forest, it may be noted that the area of forest included in national parks and other protected areas in North America (USA and Canada) was reported to be nearly 49 million ha in 1990.

The FAO Forest Resource Assessment 1990 did not distinguish between undisturbed and disturbed 'natural' forests in developing countries. However, the Forest Resource Assessment 1980 made estimates of the areas of undisturbed closed forests (primary forests and old secondary forests where there had been no logging for the last 60-80 years) and of closed forests that were included in national parks and other protected areas (thus relatively undisturbed, at least in theory). At that time, these two categories together represented 60% of the total closed forest area in the tropics, a proportion varying from 39% in tropical Asia to 59% in tropical Africa and 69% in tropical America.

These different proportions by region reflected a slower development of large-scale harvesting in tropical America compared with tropical Africa and Asia, and also the fact that, in tropical America, spontaneous colonisation did not follow in the wake of logging as systematically as in the two other regions due to lower population pressure. Although the two categories do not match the concept of 'virgin forests' of Europe and of 'old growth forests' of North America, the sum of the two categories nevertheless gives an indication of the amount of forest disturbance (or management, depending on the point of view of the observer) that existed around 1980 in the humid tropics. Although corresponding estimates for 1990 and 1995 are not available, it is likely that the share of undisturbed forests remains higher in the three tropical regions than in Europe and probably in North America too.

DEFINITION OF TREES OUTSIDE THE FOREST

Trees outside the forest are defined by default, as all trees excluded from the definition of forest and other wooded lands. Trees outside the forest are located on "other lands", mostly on farmlands and built-up areas, both in rural and urban areas. A large number of TOF consist of planted or domesticated trees. TOF include trees in agroforestry systems, orchards and small woodlots. They may grow in meadows, pastoral areas and on farms, or along rivers, canals and roadsides, or in towns, gardens and parks. Some of the land use systems include alley cropping and shifting cultivation, permanent tree cover crops (*e.g.* coffee, cocoa), windbreaks, hedgerows, home gardens and fruit-tree plantations.

Classification of trees outside the forest presents certain difficulties. There are existing classifications for agroforestry, but none applicable to all trees outside the forest. For practical reasons, the FRA 2000 definition of "forest" combines aspects of both land cover and land use. This approach creates difficulties not only for classification of forest, but also for classification of TOF.

In a study on data-gathering on TOF in Latin America, where classification was primarily based on land use criteria, separating land use and land cover aspects was found to be a main source of misinterpretation. There was a possibility of confounding coffee plantations and trees in pasture with forest, given their high density.

This clearly shows some of the problems involved in establishing a simple and reliable *a posteriori* classification.In France, the National Forest Inventory (IFN) and Teruti Land Use Study have begun to attempt coordinating classifications of trees outside the forest. The objective is eventually to use the annual Teruti data to update the IFN ten-year data, with a single national nomenclature as a possible end result.

FUNCTIONS AND CHALLENGES

In industrialised countries, farmers list shade and shelter, soil protection and improvement of the landscape and rural environment as their main reasons for growing trees. In the tropics, farmers grow woody species for food security and subsistence. Trees outside the forest are a major source of food. Livestock fodder produced by TOF can be a matter of life and death in semi-arid or mountainous areas.

Fuelwood remains the prime source of energy in developing countries, representing up to 81 per cent of the wood harvest (FAO 1999). In contrast, in the industrialised countries, fuelwood accounts for less than 10 per cent of total fuel consumption (FAO 1998). Very few studies have reported on overall fuelwood output from stands and single trees outside the forest, but agroforestry systems and orchards are known to provide a large part of the resource.

Trees outside the forest have an important ecological role. Planted trees and shrubs in fields help to check Run-off and erosion and control flooding, as well as helping to purify water and protect against wind. Trees lining rivers and streams help to maintain biodiversity, providing spawning beds for fish and shellfish and shade which reduces eutrophication.

The unique role of trees in soil protection and conservation, checking wind and water erosion and maintaining soil fertility is universally acknowledged. Also important are the cumulative benefits of trees on smallholdings to soil and water conservation, in particular in the larger context of mountain watershed management; their positive impact on climate; and their role in buffering the effects of desertification and drought.

RESULTS OF SELECTED STUDIES

In spite of the limits of data at the regional or global level, a number of local initiatives have been carried out. The approaches of the various studies differ in accordance with the purpose and scale of the analysis. Few studies use methods resembling the conventional forest inventory. Many studies rely on existing literature or estimates drawn from surveys and interviews. The quantification of products is often based on different parameters, such as estimates of global output, marketed output, observed or potential productivity or economic value. Thus the reliability of the results is uncertain.

The following are the results of some national initiatives that have assessed trees outside the forest.

In Kerala, the most densely inhabited state of India, a study estimated that of the total annual production of 14.6 million cubic metres of wood in the state, about 83 per cent was from homesteads (house compounds and farmlands), 10 per cent from estates (plantations of rubber, cardamom, coffee and tea) and only about 7 per cent from forest areas 26.6 per cent of the state area is under forest cover. Trees outside the forest met about 90 per cent of the fuelwood requirements of the state. Fuel from coconut trees alone, including both wood and non-wood materials (pruned and fallen), constituted about 70 per cent of the total fuelwood supply.

A study in Haryana State in India, an intensively cultivated state with about 3.8 per cent of its area classified as forest land but only about 2 per cent under actual forest cover showed that farm forestry (trees along farm bonds and in small patches up to 0.1 ha) accounted for 41.2 per cent of the total growing stock of wood. Multiple tree rows along roads and canals accounted for 13 per cent and 9.6 per cent, respectively; village woodlots for 24 per cent; and block plantations of less than 0.1 ha for 10.6 per cent.

In Morocco, where forest cover is less than 5 per cent of the land cover and other wooded lands only 7 per cent, nearly 20 per cent of the land may be occupied by trees outside the forest, namely as wooded pasture (84 per cent) and fruit-tree plantations (12 per cent) (Rosaceae, citrus, olives trees, palm trees, walnut trees, fig trees, almond trees).

Fruit production has an important place in the national economy. It is noteworthy that even when a forest is largely destroyed, the carob is one of the few species traditionally conserved, as it is highly appreciated by farmers for multiple purposes, providing both fodder and income from the sale of its fruit for export. However, there are no reliable data on the distribution and potential of this "forest" resource, which is of interest for farmers, herders, concessionaires and the government and distributed on agricultural and forest lands.

In the Sudan, the National Forest Inventory has undertaken a national land use inventory to provide area and volume statistics for planning at the subnational and national levels. The inventory was designed to provide preliminary estimates regarding products other than the traditional fuelwood and timber, such as the amount of gum, fruit or nuts that can be collected and the distribution of non-wood species of interest.

In Costa Rica, the Tropical Agricultural Research and Higher Education Centre (CATIE), in collaboration with Freiburg University, Germany, is developing a regional methodology for Central America to assess tree resources outside the forest.

A mix of satellite remote sensing, aerial photos and ground sampling is used to address the complexity of the resource and to allow dynamic monitoring of resources at the national and regional levels data-gathering on TOF in eight Latin American countries. None of these countries had established a database and the search for information was multisectoral. The statistics on land cover and land use gave some idea of the relative importance of trees outside the forest.

In Kenya, extensive tree planting on farmlands was promoted in the 1970s and 1980s, with land tenure security as a major incentive. There is an increasing trend of tree cover and species diversification on privately owned farms. Assuming that the present rate of increase in tree planting will continue, it was estimated that farms produced about 9.4 million cubic metres of wood in 2000 and will produce about 17.8 million cubic metres in 2020.

Their share of the total wood produced in the medium- and high-potential districts was projected to increase to 80 per cent in 2020. Indeed, while natural stands of trees have declined there has been a corresponding increase in tree planting in much of the densely populated plateaus of Kenya. As natural forests are reduced or become inaccessible, agroforestry systems help people to diversify production and income and to protect themselves from shortages of fuel and wood.

In Bangladesh, natural forest formations cover less than 6 per cent of the country and the population growth rate is extremely high. An inventory of homestead/village forests in the country indicated that trees outside the forest constitute a vital resource for local populations, providing food, fodder and fuelwood. The sampling method was based on dual village/household sampling with an agro-ecological and administrative sampling base. Rural Bangladesh was divided into six major regions considered as agro-ecological strata, each subdivided into *thanas* (administrative entities, subdistricts). The households making up the sampling units were chosen at random from a number of villages.

The inventory sampled data on palm trees and cane as well as trees, bamboo and thickets. The results, expressed per stratum and per inhabitant, provide volumetric data for fuelwood and sawnwood and species data for total amounts under and over 20 cm. This inventory was apparently the first to nationwide assessment of trees outside the classified forests in Bangladesh.

METHODS AND TOOLS FOR FUTURE ASSESSMENTS

A priority challenge of future assessments is to know the state and dynamics of all tree resources both in and outside the forest. A country embarking on a planning exercise cannot confine itself solely to the trees within its forests, especially when its wood resources appear to be insufficient.

The choice of tools and methods used to describe or assess trees outside the forest depends on the scale of analysis, kind of data and degree of exactitude desired. The tools used are not generally specific or new; rather, they are combined and implemented in original ways.

The inventory in Bangladesh described above is one of the numerous examples of methods developed for gathering data on TOF. The Bangladesh study gives evidence of the adaptations needed to bring conventional forest inventory procedures in line with the specific nature of this resource.

In some aspects - *e.g.* structure, spatial distribution and extent of area cover - trees outside the forest are more difficult to assess than forest formations. The assessment of TOF does not lend itself to the potential cost savings associated with expanded uses of remote sensing technology. Remote sensing by satellite presents more difficulties for assessing TOF resources than for assessing attributes such as forest area. However, satellite data do allow a region to be stratified on the basis of ecological criteria and land cover, providing the basis for a good working document for more specific work in the future. The most commonly used remote sensing technology for TOF resources is aerial photography, which can be used to describe spatial distribution and to distinguish

TOF cover classifications, providing the appropriate scale is chosen. However, high costs prohibit widespread use of aerial photography for TOF assessments in most countries. The new 1 m resolution satellite sensors represent a possible future alternative to aerial photography.

Some TOF field inventories are modelled on forest inventory methods and keep to biological and physical criteria; others emphasize social aspects, choosing villages as the sampling units. For measurements on the ground, sampling arrangements designed for forest stands may not be the most effective arrangements for trees. Less traditional sampling plans which would theoretically be better suited to this resource should be tested on various categories of TOF, especially those covering fairly large areas.

Studies of the social and economic benefits or impacts of TOF often rely on household surveys, interviews or standardised appraisals such as rapid or participatory rural appraisal. The integration of the last two approaches - biophysical inventory and socio-economic analysis - is not simple and calls for caution given the great variety of social situations that are only meaningful in the local context.

Environmental benefits or impacts of TOF might be indirectly assessed by linking measurable indicators, such as the number and type of trees, with environmental variables such as water quality or erosion. In an urban setting, tree cover might have direct impact on the ambient temperature. Measuring the environmental impact of tree management is an issue for all natural resource planning or management operations.

Assessment of trees outside the forest requires geographical, ecological, biophysical, social and economic data. However, this implies that an important amount of information will have to be carefully processed. The diversity of end-uses for this information, including land use planning and analysis based on inventory, will need to be considered in data assembly and processing and in the presentation of results.

It is important to know the status of trees outside the forest at any given moment, but it is even more essential to be able to trace patterns of change over time in the same area. The two most commonly used approaches have been comparison of aerial photos taken at sufficiently long intervals and surveys among villagers/managers combined with field inventories.

Some countries, such as France and the United Kingdom, have undertaken periodic inventories based on the establishment of permanent plots linked to permanent forest inventories. However, the high cost of this type of operation limits the number of countries able to adopt it. India and Bangladesh are now experimenting with options for the future.

The current trend towards decentralised authority in land use planning suggests the importance of carrying out assessments at the local level, where the geographical, historical and socio-economic context is relatively harmonious.

A minimum number of common rules concerning methods and arrangements is necessary, however, if the data are to be comparable at the country level. Certainly, the technical side of assessing trees outside the forest is complex and more research is needed to better pinpoint the resource.

FOREST TYPES AND REGIONS

FOREST TYPES

The forest flora of India is rich and varied both in composition and value. The number of species is estimated at 30,000, although the vast majority have no economic value. The distribution of forests and forest types as elsewhere, depends on climatic, edaphic (including relief) and biotic factors acting in combination. Hence, no one of these factors considered alone can give a satisfactory clue to the nature of the forest in any particular locality. Indeed, a forest itself is often a better index of climate, soil and certain biotic conditions than vice versa.

THE CLIMATIC FACTOR

Marked climatic contrasts reflect pronounced differences in the relief of India. The mercury rises above 120°F. at Jacobabad in Sind and falls as low as — 49°F. at Dras in Kashmir. A climb of only a few miles brings one from hot, moist marshes (tarai) along the outer foothills of the Himalayas to the temperate climate of the former British hill stations (ca. 6,500 feet) and thence to arctic cold above the snow line (ca. 12,000 feet). The contrasts in rainfall are even greater and have a profound effect on the vegetation, the extremes ranging from an annual mean of only 3 inches in Upper Sind to nearly 40 feet in the Khasi Hills.

Three seasons may be distinguished: winter (November-February), summer (March-June) and the rains (July-October). During the winter, winds of continental origin blow from the land Towards the sea, in general from the northeast and northwest. This "northeast monsoon" becomes fully established by January, bringing cool, dry weather, although local disturbances cause occasional showers in northern India and in the eastern Deccan. The winter rainfall is heaviest in the northwest. As summer advances, the wind direction is reversed. The winds now blow from sea to land, culminating in the "southwest monsoon," which sets in with the rainy season. The summer is characterised by progressively rising temperature and falling atmospheric pressure. Temperature maxima range from 100°F. in March in the Deccan, to 110°F. in April across central India and 115°F. in May and June throughout the IndoGangetic plains, reaching extremes of 120°F. in the northwestern desert.

Coinciding with the area of lowest pressure, this blazing furnace in the northwest induces indrafts of the monsoon from the Arabian Sea and Bay of

Bengal.The bulk of the rainfall is brought by the southwest monsoon during the four months July-October. One great current of the monsoon bursts against the Western Ghats and flows over the Deccan, another sweeps up the Bay of Bengal and, combined, they flood the eastern Himalayas and outlying hills to the south. July and August are the rainiest months here. Northwestern India, away from the main paths of the southwest monsoon, gets the lowest rainfall in the country. For the purposes of the following description of its forest types, India will be divided into six major physical regions designated somewhat arbitrarily:

- The Deccan,
- The Indus Basin,
- The Gangetic Basin,
- The Himalayas,
- The Seacoasts,
- The Andaman Islands.

While these provide a convenient means of classification, it should be borne in mind that they are based on criteria of surface relief and not of phytogeography. Thus, more than one region may include representatives of a single forest type and divers forest types are found within each region. The estuarine vegetation in the delta of the Ganges (or Ganga), for example, is like that of the deltas of the eastern coast of the Deccan; the evergreens of the Malabar Coast have far more in common with those of the northeastern Himalayas than with the vegetation of the rest of the Deccan.

MAIN FOREST TYPES

In describing the forests of these regions we shall follow Champion's classification in broad outline, omitting details. Five main forest types and several transitional, intermediate and subtypes will be distinguished. The main types are:

- *Rain Forest:* Annual rainfall in general 80 to 120 inches; dense, tall evergreens.
- *Monsoon Forest:* Rainfall 40 to 80 inches; dominants deciduous; evergreens form the second story.
- *Dry Forest:* Rainfall 30 to 40 inches; deciduous; no evergreens.,
- *Xerophytic Forest:* Rainfall 10 to 30 inches; low, stunted, thorny, deciduous species and fleshy evergreens.
- *Mangrove Forest: R*ainfall varies; both evergreen and deciduous species.

Changes in the floristic composition of a given type brought about by edaphic and biotic factors and by rainfall and other climatic factors will be pointed out. For example, coarse, friable, porous soils require a much higher rainfall to support a rain forest than do soils with higher moisture retentivity. The presence of alkalis and salts may retard and even eliminate tree growth altogether, despite high moisture content. Human exploitation, fires, grazing and lopping often act as limiting factors, affecting the character of the vegetation independently of the rainfall. The range of temperature and, more particularly, minimum temperatures, sharp frosts and snowfall have a profound influence. In the Himalayas the vegetation of the northern latitudes of the temperate zone is reproduced.

The vegetation in the Deccan plateau, with its equable climate, differs from that of the Indo-Gangetic plains, which exhibit a continental range of temperatures, both seasonal and diurnal. The Nilgiri Hills, in the south, have a vegetation unlike that of the Himalayas at the same range of elevation. A distinction must often be made between tropical and Himalayan varieties of the same main forest type. The similarities between a given type of forest occurring in the Deccan and its counterpart in the north, however, may be fully as striking as the differences. The vegetation of the rainless Thar Desert is reminiscent of that of alpine wastes both in form and appearance.

THE FOREST REGIONS

THE DECCAN

The term "Deccan" should be here understood as embracing the whole of peninsular India and of the Central Highlands, with the exception of the northern and northwestern slopes of the latter, which will be considered as parts of the Gangetic and Indus basins, respectively.

The Deccan plateau rises above the Indo-Gangetic plains in a series of irregular terraces. Its principal rocks are of archean origin, comprising granites, gneisses and schists, with basaltic lavas of more recent age capping them, particularly in the northwest. Decomposition of these lavas has given rise to the rich "black cotton soils." The plateau is buttressed on the north by the Aravalli, Vindhya, Satpura and Kaimur hills, with characteristically flat tops. The Aravallis (ca. 5,500 feet), an extension of the plateau to the northwest, form an isolated, ancient and much worndown range flanking the desert of Rajputana on the east. Farther south, the eastern and western flanks of the plateau are fringed by the Ghats, which converge to an apex in the south. With an average elevation of about 1,500 feet, the plateau surface slopes down eastward from an altitude of about 3,000 feet in Mysore. The Western Ghats begin near the Tapti River and increase in elevation southward, reaching nearly 9,000 feet in the Nilgiri Hills. The Eastern Ghats average about 1,500 feet in altitude.

In considering the forest types of the Deccan, we may conveniently divide the region as a whole, exclusive of the seacoast, into three lesser regions: the West Coast, the Southeast and the Central Plateau.

THE WEST COAST

The region comprising the Malabar and Konkan coasts, the Western Ghats and the Nilgiri and Cardamom hills has the most luxuriant vegetation in the Deccan, owing largely to the heavy rainfall and to a soil rendered rich by the decomposition of a vast mass of leaf litter. The southwest monsoon floods the entire region. The mean annual rainfall increases with elevation on the Western Ghats; thus, while Bombay at sea level receives only 74 inches, Matheran at an altitude of nearly 2,000 feet some 25 miles to the east receives 200 inches. The higher altitudes of the southern portion of the Western Ghats also helps explain why the rainfall on the Malabar Coast (some 130 inches) is nearly twice as heavy as that on the Konkan Coast to the north (75 inches).

Of the main forest types, Dry Forest and Xerophytic Forest are not found in this region. The Mangrove Forest will be mentioned under another heading. Besides Rain Forest in the strict sense, there are two intermediate types: Shola and Semi-Evergreen Forest.

Rain Forest

Forests of this type are generally met with along the Malabar Coast and throughout the Western Ghats up to altitudes of 1,000 to 1,500 feet, wherever the mean annual rainfall is above 100 inches (with 120 inches as the optimum). In these areas the mean annual temperature varies from 80° F. to 100° F., the January minimum seldom goes below 65° F. and the mean annual humidity is about 80 per cent (*e.g.*, at Cochin). Under such humid tropical conditions a bewildering variety of species flourish, their only common characteristic being their evergreen nature.

These forests present several ill-defined stories, or tiers, the highest containing lofty trees, often buttressed, which reach a height of 150 feet and have girths of 15 feet or more. The lower tiers consist of an impenetrable mass of evergreens covered with a luxuriant growth of epiphytes (mosses, ferns, orchids) and climbers.

The most typical genera are Dipterocarps and Hopeas, although the former are the less frequent and may even be locally absent.

Species commonly met with are: Dipterocarpus indicus, Hopea parviflora,Hopea wightiana,Calophyllum tomentosum, Cullenia excelsa,Palaquium elliptica,Dysoxylum malabaricum, Cedrela toona,Vateria indica,Canarium strictum,Tetrameles nudiflora, Mesua ferrea,Mangifera indica,Sterculia alata, Artocarpus spp., Acrocarpus fraxinifolius and many other Myrtaceae and Lauraceae. The undergrowth, conspicuously free from grasses, is usually a tangled mass of cane, congested and creeping bamboo and palms.

Sholas. Above the rain forests, related forests of a special intermediate type, locally known as sholas, are found up to about 4,500 feet in the valleys and on the slopes of the Nilgiri, Anaimalai, Palni and other hills of southernmost

India. At relatively lower elevations, these sholas usually consist of Mesua ferrea,Vitex altissima,Artocarpus lakoocha and Dysoxylum malabaricum, often with Strobilanthes and Selaginella as a ground cover and Arundinaria spp. and Oxtenanthera spp. as the commonest bamboos.

At higher altitudes, the floristic composition changes to include such subtropical and temperate species as Michelia nilagirica, Ternstroemia japonica,Eurya japonica,Gordonia obtusa,Eugenia spp., Ilex spp., Meliosma wightii,Photinia notoniana,Cinnamomum wightii, Euonymus crenulatus.

Here, owing to comparatively rigorous weather conditions, the tree growth suffers in height and the crowns are more rounded. Mention must also be made of artificial plantations of Australian eucalyptus and of South African wattles. Exotic pines and cypress have also been introduced in the Nilgiri and Anaimalai hills. Farther north, on the Western Ghat slopes — as for example, at Mahableshwar — sholas are found in which the representative species are Eugenia jambolana, Actinodaphne hookeri,Canthium didymum,Terminalia chebula,Olea dioica,Strobilanthes callosus and Sideroxylon tomentosum.

Semi-Evergreen Forest

Along gentle slopes, where the mean annual rainfall is between about 80 and 100 inches, the evergreen nature of tile rain forest is further modified by the increasing association of deciduous species, typified by Terminalia paniculata,Xylia dolabriformis,Stereospermum chelonoides,Trewia polycarpa, Eugenia spp., Diospyros spp., Lagerstroemia lanceolata, Schleichera trijuga,Vateria indica, with Bambusa arundinacea as the representative bamboo in the lower story. The middle canopy is usually composed of evergreens, Myrtaceae and Lauraceae and the ground cover includes various species of Rubiaceae and Acanthaceae. Canes and climbers persist.

Monsoon Forest

The semi-evergreens merge imperceptibly into a moist, deciduous "monsoon forest," comprising comparatively few species but some of considerable economic importance, such as teak, rosewood, Xylia spp., Terminalia spp., Lagerstroemia lanceolata, Pterocarpus marsupium,Adina cordifolia, Stephegyne parvifolia and Grewia leptopetala, which occur in more or less gregarious associations. Schleichera and Careya are usually present in the second story. Bambusa arundinacea,Dendrocalamus strictus and Oxytenanthera monostigma are the representative bamboos, although often locally absent. The undergrowth may consist of near-evergreen shrubs but more frequently of grass. This monsoon forest occurs throughout the western Deccan in regions of 40 to 80 inches of rainfall, more particularly above the Palghat Gap. The mean annual temperature varies from 75° F. to 80° F., with the January mean often below 60° F. The trees, which reach a height of 100 to 120 feet, are generally leafless during March and April, putting on new foliage somewhat later. Along the Konkan CoastTerminalia spp. lend their name to the forest wherein they abound, in which teak may or may not be present locally.

THE SOUTHEAST

The southeastern subregion of the Deccan comprises the Carnatic, the Eastern Ghats and the bulk of the plateau south of the Godavari River. During the period of the southwest monsoon, the region lies in the rain shadow of the Western Ghats. The rainfall, generally associated with the northeast monsoon, when the rest of India is dry, is relatively low and extremely variable. The annual precipitation seldom exceeds 40 inches, the average varying between 20 and 30 inches. Of the main forest types, only the two drier ones are represented here.

Xerophytic Forest

In the Carnatic and farther south in the lee of the higher Western Ghats the vegetation is of a dry, thorny, stunted type. The most representative genus is *Acacia,* with low (20 to 30 feet) crowns. The ground is usually bare, with thorny shrubs dotted here and there and light grass appearing during periods of rain. The mean annual rainfall varies between 20 and 30 inches (with about eight months with less than 2 inches) and the mean annual humidity between 50 and 60 per cent. The mean annual temperature is about 78° F., the mean maximum about 105v F. and the hottest month is May (85° F. to 90° F.).

Dry Forest

The xerophytic thorn forest of the Carnatic gradually merges into a dry deciduous forest, where the mean rainfall is between about 35 and 40 inches, the mean temperature is close to 75° F. and the mean annual humidity is close to 60 per cent. This forest forms an intermediate stage between the monsoon and xerophytic forests, having characteristics of both. The trees are seldom more than 75 feet high. More species thrive under these conditions than in the xerophytic forest and the dry forest includes the most valuable tree of southern India, sandalwood (*Santalum album*). Gregariousness is not common, unless artificially brought about. This type of forest usually occurs on shallow, sandy soils and laterites and in places where the rainfall is relatively heavy but rapidly drains off.

In the northern part of the region the dry forest is fairly dense. The representative species are teak (Tectona grandis) and associated Terminalia spp. and Anogeissus latifolia. Other species include Dillenia pentagyna,Kydia calycina, Lagerstroemia lanceolata, Adina cordifolia, Bassia latifolia, Buchanania latifolia,Grewia asiatica, Odina wodier, Pterocarpus marsupium, Bombax malabaricum, Santalum album, Butea frondosa, Bridelia retusa, Diospyros melanoxylon and Phyllanthus emblica. The chief bamboo is the well-known commercial variety, Dendrocalamus strictus. Grasses are plentiful, including Ischoemum angustifolium, Flemingia spp., Millettia auriculata,Lantana,Carissa spinarum, etc.

Farther south, teak is less common and in many places absent, owing, perhaps, to past human exploitation. Terminalia spp. and Anogeissus spp. are associated with Acacia catechu, Chloroxylon swietenia,Cleistanthus collinus,Hardwickia binata, Boswellia serrata and Soymida febrifuga. The ground cover is composed of grasses, usually Andropogon spp., Panicum spp. and Aristida spp.

THE CENTRAL PLATEAU

For present purposes the Central Plateau may be regarded as comprising the entire region between the Godavari River and the Bay of Bengal on the south and southeast, the Western Ghats and Aravalli Hills on the west and northwest and the northern slopes of the Central Highlands on the north and northeast.

North of the source of the Godavari, the Western Ghats become progressively lower and finally disappear and, hence, no longer constitute an effective barrier to the southwest monsoon. The hill ranges of the Central Highlands run in an east-west direction, which allows a more even distribution of rainfall over this part of India than farther south. The climate becomes more continental, with marked extremes, as one goes northward. Of the major forest types, xerophytic, dry and monsoon forests are represented on the Central Plateau. At higher elevations the shola type, which is intermediate between the monsoon and rain forests, is also met with.

Xerophytic Forest (rainfall, less than 25 inches). Here, the general pattern of xerophytic vegetation found in the Carnatic is closely repeated. Acacia leucophloea, Balanites roxburghii,Cordia myxa, Zizyphus jujuba and Calotropis procera are characteristic species.

Dry Forest (rainfall, 35 to 40 inches). This is met with throughout the region, with low-quality teak much more frequent than in the corresponding dry forests farther south. Here, also, teak may be locally absent, suggesting human exploitation in the past. Any one of such species as Hardwickia binata,Boswellia serrata,Butea frondosa and Anogeissus pendula may be locally preponderant. Babul (Acacia arabica) forests grow more particularly on the rich black cotton soils, notably in Berar and in the vicinity of Poona.

Monsoon Forest (rainfall, 40 to 80 inches). From the Maikal Hills eastward and especially on the Chota Nagpur plateau, there is a copious rainfall due to the convergence of the two great currents of the southwest monsoon. Hence, the vegetation is rich and clearly assumes the form of monsoon forest. Where teak is more or less gregarious, the forest is known as "moist teak forest," and is not unlike the Terminalia forest of the West Coast region of the Deccan. Teak in this region of the Central Plateau is associated with Terminalia tomentosa,Adina cordifolia,Pterocarpus marsupium,Dalbergia latifolia,Odina wodier,Lagerstroemia parviflora,Schleichera trijuga,Ougeinia dalbergioides and Phyllanthus emblica, with Dendrocalamus strictus as the common bamboo. Helicteres isora covers the ground, along with Millettia auriculata and Bauhinia vahlii. Where sal (Shorea robusta) predominates, in the so-called "moist sal" or "peninsular sal" forests, the associated species are Pterocarpus marsupium,Terminalia chebula (and T. tomentosa), Anogeissus latifolia,Bassia latifolia,Eugenia jambolana,Phyllanthus emblica and Indigofera pulchella, with Phoenix acaulis,Flemingia chappar and Anthistiria ciliata L. as the ground cover. Sholas (rainfall, about 100 inches). On the upper levels of the Maikal Hills (*e.g.*, near Pachmarhi) and of the Chota Nagpur plateau (*e.g.*, near Parasnath), the sholas of the Western Ghats are repeated, with slight modifications due to soils derived from basaltic lavas. Pterospermum acerifolium,Phoenix robusta and Clematis nutans are representative species.

THE INDUS BASIN

The Indus basin forms a vast area of low rainfall and high summer temperatures. The rainfall increases in a generally northeastward direction, for the clouds borne by the southwest monsoon current from the Arabian Sea tend to pass over the Thar Desert without dropping their moisture. In the Punjab the rain brought by this current is to some extent supplemented by rain from the Bay of Bengal. In the North-West Frontier region, however, the force of both monsoon currents is feeble and, consequently, cold-weather rains account for more than half of the annual rainfall.

Recent excavations at Mohenjo-Daro in Sind and the discovery of elephant and rhinoceros remains along the right bank of the Indus point to a time when there must have been moist forests in this part of India and the chroniclers of Alexander's invasion described the Larkana area in Sind as a "garden." Conditions are very different today. Two main forest types may be distinguished: dry forest and xerophytic forest.

Dry Forest

This type occurs in the northern Punjab where the mean annual rainfall is between 35 and 40 inches, the mean annual temperature between 75° and 80° F., the mean annual minimum and maximum are about 35° and 100° F., respectively and both the diurnal and the seasonal ranges are wide. Owing to these severe conditions the trees seldom exceed 50 feet in height.

In summer the lack of leaves gives a desolate look to these forests, but they turn bright green at the break of the monsoon. Among the species met with, Anogeissus latifolia is typical, its usual associates being Buchanania latifolia,Sterculia spp., Bauhinia spp., Acacia catechu,Aegle marmelos,Limonia acidissima,Ehretia laevis, Kydia calycina, Ougeinia dalbergioides and Terminalia tomentosa. Mallotus spp. and Nyctanthes occur in the lower story. Dendrocalamus strictus is the usual bamboo. The undergrowth is composed of Woodfordia floribunda,Indigofera spp., Carissa spinarum, Adhatoda vasica, Dodonaea viscosa,Bauhinia vahlii and Acacia caesia.

Grasses are common. But for certain differences in emphasis, the dry deciduous type of the South is closely reproduced. Owing to the more pronounced climatic extremes, however, the teak, *Chloroxylon,Soyrnida* and *Cleistanthus* met with in the dry forests of the Deccan are conspicuously absent and the *Dalbergia latifolia* is replaced by its hardier cousin, *Dalbergia sissoo.*

Xerophytic Forest

Around the margins of the Indus Plain,' wherever the mean annual rainfall is between 10 to 25 inches — as, notably, in Cutch and Kathiawar, along the base of the Kirthar and Sulaiman ranges, on the Salt Range, in the Sirhind Gap, on the Delhi Ridge and along the lower western slopes of the Aravalli Hills— a scrub, seldom more than 3 feet high, is found, corresponding to the xerophytic forest of the Deccan. Here temperature conditions are comparable to those of the dry forest area of the Punjab.

Various species of acacia abound, the chief being *Acacia arabica* and *A. leucophloea* and are associated with *Prosopis spicigera* and the exotic *Prosopis juliflora,* which has come to occupy the Delhi Ridge and parts of Rajputana. *Acacia catechu* is found in coniderable quantity in Cutch. In the southern Aravallis *Anogeissus pendula* occurs in fairly homogeneous patches.

Farther north, typical species are *Salvadora oleoides,Tamarix articulata,Zizyphus jujuba, Grewia populifolia* and *Ehretia laevis.* The undergrowth consists of *Capparis aphylla* (and *C. spinosa*), *Asclepias* and *Calotropis,* with the fleshy *Suaeda* and *Salsola* growing on saline soils and *Calligonum polygonoides* on sandy soils.

Light grass flourishes only during the rains. On the floodplain of the Indus, which is periodically inundated and in the regions with slightly higher rainfall, *Acacia arabica,* a species that adapts itself admirably to xerophytic conditions, occupies the ground, usually in association with *Populus euphratica.Dalbergia sissoo* and *Acacia catechu* are the first colonizers of the alluvial deposits along the river banks in their Himalayan reaches.

Lower down (in southern Punjab and Sind), fresh alluvial deposits are first occupied by pure associations of *Tamarix dioica* and *Populus euphratica,* which give place in the course of time to *Acacia arabica* and *Dalbergia sissoo,* though the first two often persist in the understory. In the Thar Desert the xerophytic vegetation takes a more retrograde form, with fleshy evergreen *Euphorbias* as the most characteristic species.

BENEFITS OF URBAN FORESTS AND TREES

Trees in the urban environment contribute significantly to the aesthetic appeal of cities, helping to maintain the psychological health of the inhabitants. Besides the aesthetic and environmental aspects, urban forestry is also of importance in helping populations poor in resources to meet basic needs. Urban plantings of fruit trees to provide firewood is one way of utilizing obvious benefits from the available natural resources, as done in Kampala, the capital of Uganda. Research has shown that urban trees benefit communities economically, socially and environmentally. The size, structure and condition of the urban forest directly affects the amount of benefits provided by the urban vegetation. Nevertheless, the social and environmental benefits provided by the urban forest are often largely ignored in land-use planning.

ECONOMIC IMPACT

Though trees are not usually thought of as economic resources, the presence of trees has an impact on different economic factors, including the level of real-estate prices, economic investments and employment. Results of recent Finnish studies show that the benefits provided by urban forests are reflected in property prices, and that environmental variables, such as proximity to wooded recreation areas and water courses as well as an increasing proportion of total forested area in the housing district, had a positive influence on apartment prices. The presence of trees may influence property values positively by as much as 20%, with an average increase of 5-10%.

Trees also provide different external environmental benefits. Costs for heating and cooling can be reduced by appropriate use of vegetation. Energy reductions for individual buildings have been shown to range from 5 to 15% for heating and 10-50% for cooling. Especially in areas with high summer temperatures, trees are important providers of shade. In areas with cool winters, it is of importance to use deciduous trees, as the loss of leaves in winter allows the sun to heat the house. Proper arrangements of vegetation around buildings can also reduce wind velocity and thereby reduce the heat loss from buildings.

RECREATIONAL USE OF GREEN AREAS

Several studies have shown the importance of urban forests for participation in outdoor recreation activities. In a Danish national survey it was found that around two-thirds of all forest visits take place in the forest situated nearest the housing area. Danish and Swedish studies have also showed that the use of green areas by people is associated with three highly esteemed values: (i) a high degree of natural elements, (ii) fresh air and (iii) the possibility of solitude.

In general, people use green areas less than they would like. According to Swedish studies, distance and the fear of assault are the most prevalent reasons for people not visiting green areas. When the distance to the park exceeds 300 m, one person in four postpones a daily visit. As many as 56% refrain from regular walks in the park when the distance increases to 500m. American studies indicate that the removal of vegetation at strategic points can reduce the risk of assault, although this should not be the reason for removing all the vegetation.

Psychological Aspects

City life is stressful but research shows that urban green areas have a beneficial influence on the health and well-being of the urban population. Studies indicate that visits to green areas can counteract stress, renew vital energy and speed healing processes.

In Sweden, Grahn (1989) conducted extensive studies on the significance of parks to different groups of the population. For instance, he persuaded 40 schools, hospitals, sports associations, cultural associations and day-care centres to keep diaries of their outdoor activities for 1 year. For example, the diaries show that periods spent outside had an actual medicinal value for patients and residents of hospitals, old people's homes and homes for the sick. People became happier, slept better, needed less medication, were less restless and far more talkative.

Ulrich (1984) showed that hospitalized patients recovered faster when they had a view through a window, allowing them to see trees. Ulrich *et al.* (1991) showed a gory film on industrial accidents to 120 people. Half the people were then shown a nature film, whereas the other half were shown a film on the city, with sequences of buildings and traffic. The subjects' heart beat, muscular tension and blood pressure were monitored throughout. All subjects exhibited strong signs of stress during the first film on industrial accidents. The stress levels of the subjects that then watched the nature film returned to a normal level after 4-6 min, whereas the half that watched the film on buildings and traffic continued to exhibit high stress levels.

Kaplan and Kaplan (1989) have formulated a theory on the interaction between human attention and the surroundings. This theory distinguishes between spontaneous attention and conscious attention. Spontaneous attention demands no effort and occurs without premeditation.

Conscious attention demands energy and leads to psychological exhaustion in the long term. Rapid movements, strong colours, sudden noises and strong odours are typical stimuli that demand conscious attention. These signals have a powerful effect on our attention, consciously and unconsciously, because they are perceived as potential dangers to which we should react. This means that urban living, with fast vehicles, flashing neon signs and strong colours, causes constant stress.

Kaplan and Kaplan's research indicates that vegetation and nature reinforce our spontaneous attention, allow our sensory apparatus to relax and infuse us with fresh energy. Visits to green areas bring relaxation and sharpen our concentration, since we only need to use our spontaneous attention. At the same time, we get fresh air and sunlight, which have significance for our diurnal and annual rhythms.

Environmental Education

The change and the continuity in nature provides not only a sense of time but also a sense of security and confidence, through the predictable change of seasons and the repetition of natural processes. The urban trees and forests play a role in educating young people and children in understanding the basic processes in nature and the complexities of the environment, either informally during play or as part of the school curriculum. Playing in nature or natural settings also helps children to develop their motoric senses and to regain and experience the human connection to nature through their imagination. Experiences with environmental education in the USA show that access to even a small area of the urban forest is of importance when learning about natural processes.

In England in 1985 a research project supported by the Department of Education and Science, the Countryside Commission and a consortium of local authorities resulted in the organisation Learning through Landscapes (LTL). LTL was founded on the recognition of the impact that the physical surroundings has on children. LTL helps schools to design schoolyards that improve the quality of the environment and helps to create additional resources for the formal curriculum, such as the promotion of more effective teaching and learning in the outdoor classroom.The English Community Forests are examples of integrated programmes in which environmental education is integrated in social programmes by well-coordinated planning. The forests are charged with running and coordinating social programmes, involving information, consultation, participation, art, education, sport, recreation and cultural activities.

Community Involvement

Other benefits of urban forestry are those provided by involving people in planning, planting and caring for the trees and urban forests in their own locality. By promoting social interaction and strengthening local pride and identity, local public engagement benefits management greatly, encouraging people to act protectively towards their local forests and providing a potentially huge volunteer workforce.

The possibilities for creating public awareness in any town are endless. In the USA there is a tradition of involving the public in all processes, from the policy-making to the care and maintenance of the urban forest. Through various types of tree programmes, public awareness has proved to be a valuable tool in the justification of tree activities because it develops a public interest group that helps to lobby for support. The projects that involve citizens have various aspects in common, such as planting, preserving, pruning, maintenance, public education, etc.. Examples of initiatives that enhance public awareness and engagement in a municipal tree programme are the Tree Boards, municipal non-profit groups with participants from a wide range of professions. The close cooperation between professionals and non-professionals provides direct communication of new ideas and research and helps to solve associated problems as experienced in a practical situation. Another way of involving the public is to encourage houseowners to adopt a tree, by taking part in the care and maintenance of a tree in their neighbourhood.

TREES IN THE URBAN ENVIRONMENT

Trees in the urban environment are often referred to as the urban forest, comprising trees in civic woodlands, parks and the street. Earlier, urban trees were mainly regarded as aesthetic elements, whereas today they are recognised as having a positive impact on the environment as well as providing economic and social benefits. Monetary evaluations reflect the various benefits arising from the urban forest, covering such aspects as reduction of pollution and energy use, environmental amelioration, savings in public health care and increase in economic investment. Hence, the value of the urban forest is being increasingly recognised as a vital component in the maintenance of a sustainable urban environment in cities around the world. Meanwhile, the population living in urban areas has increased rapidly since the 1950s and the lack of space makes it tempting to use green areas for infrastructure and buildings.

Collins (1997) emphasizes that urban forestry, the planning, management and maintenance of the urban forest, is more closely aligned to traditional forestry than might be immediately apparent. As implicit in the term, the principle of sustained yield has been adapted to the urban environment, applying rural land-use forestry to the management of the urban forest. Hence, the overall objective in urban forestry is not that of timber production but, through a balanced structure of age and species, a sustained production of environmental, social and economic benefits. These social benefits also accrue if local communities are encouraged to contribute to their own environment by promoting projects and activities involving local residents. These kinds of projects often prove effective in promoting social interaction and lead to increased local involvement in other aspects as well.

The planning and management of a healthy urban forest requires that coordinated strategies are agreed between the professions more or less directly involved, such as planners, landscape architects, arboriculturists, foresters, engineers, legislators, developers and utility managers. Management of the urban forest resource is moving towards the achievement of a healthy, well-distributed urban forest, which is increasingly becoming an essential and integral part of the urban infrastructure.

Trees in the urban environment have been defined in several ways, although 'urban forestry and arboriculture' is probably the most used term in relation to trees in or near the urban environment. Many different research disciplines are involved in the fields of arboriculture and urban forestry. Harris (1992) defines arboriculture as being 'primarily concerned with the planting and care of trees and more peripherally concerned with shrubs and woody vines and groundcover plants'. However, many people consider urban forestry to consist primarily of two types of planting: urban forests and urban trees. Also, definitions of urban green areas and urban forestry vary significantly throughout the world. There seem to be different opinions as to what urban forestry covers, depending on whether professionals have a background within or outside forestry or have experience of working in the USA or other parts of the world such as Europe.

In many cases foresters have argued that urban forestry concerns 'forestry in urban areas'. Volk (1986) stated that:

> Green areas such as tree lined streets, cemeteries, and parks, which are not predominated by trees and as such do not come under forestry influence should be excluded from consideration [of being considered urban forestry]. On the other hand, park forests which come under forestry supervision, should be included in our considerations. The fear of involvement of foresters in urban plantings has been described by Chambers (1987): 'Misguided resistance to the concept [of urban forestry] is still often mistakenly assumed to imply the take over of public open space and existing amenity trees for timber production.' Americans tends to look at urban forestry as the 'management of trees in urban areas on larger than an individual basis'. In Europe this broad concept now seems to be accepted on a wider basis. Urban forest stands are now often considered as resources where an economic yield is not required, and traditional forestry practices are combined with both aesthetic and recreational considerations.

Costello (1993) suggested that urban forestry be defined as 'the management of trees in urban areas', including single trees. In this definition, 'management' is described as the planning, planting and care of trees; 'trees' as individuals, small groups, larger stands (*e.g.* green belts) and remnant forests; and 'urban areas' as those areas where people live and work. The location of urban forests can be anywhere, from urban settings to the countryside, as long as there are human structures on the site; these structures can be related to ecological functions, protection, merchandise or tourism and recreation.

Using this concept, we categorize trees in the urban environment as both urban forests and urban trees. Urban forests can be defined by their placement in or near urban areas and by their multifunctional aspects, giving shade, amenity values, etc. Therefore, we define urban forestry as the establishment, management, planning and design of trees and forest stands with amenity values, situated in or near urban areas.Trees in the urban environment may be divided into three different types: trees in urban woodlands, park trees and street trees. These types of plantings differ distinctively in several ways. Most importantly, they have significantly different growing conditions, which means that they have different needs with regard to planning, management and maintenance. The methods used and the research problems related to urban green areas are common throughout the world due to the fact that urban 'greening' concerns not only natural spheres (such as trees, growing conditions, etc.) but also the environment in close proximity to human populations.

URBAN GROWING CONDITIONS

Urban growing conditions differ significantly from those in the rural landscape, and produce difficulties as a result of both above- and below-ground influences.

Stress Factors

The harsh soil and air conditions that exist in urban planting are problems that do not play the same role in landscape planting. Growing conditions may also be difficult due to shading effects, recreational users, etc.. The modified urban mesoclimate affects the quantity of contaminants in urban areas, which is raised by a factor of around 25. In general, the average lifespan of a newly planted street tree may be as low as 10-15 years.

During the last 30-40 years, the vitality of street trees has fallen drastically. Heavier traffic patterns have increased demands for road construction, which consequently has changed the growing conditions of many roadside trees. Also, pollution from traffic has a highly detrimental impact on street trees. The fact that 50% of the trees planted in an urban environment die within the first year emphasizes this point. Nowak *et al.* (1990) found that 34% of 480 trees died within 2 years of planting, while Miller and Miller (1991) found that the mortality rate was 25-50% for a number of species planted in Wisconsin, USA.

Temperature extremes can occur, especially where trees are widely spaced and where heat is reflected from hard surfaces. Harris (1992) described that, occasionally, tree limbs up to 0.6 m and trunks up to 1.2 m in diameter break and fall during hot calm summer and autumn afternoons and subsequent evenings. Roots are more sensitive to temperature extremes than the tops of plants.

Wind speed will vary according to the shape and height of buildings. Areas with tall buildings will usually be relatively cool in summer due to shading effects, and warmer in winter due to wind-protection effects. On the other hand, winds are more variable and more extreme at exposed corners of tall isolated buildings. Buildings deflect strong winds downwards and concentrate their force at the base and corners of buildings, forming 'wind tunnels'. Trees planted in these exposed gaps may suffer scorched leaves and shoots, which lead to a stunted canopy, especially on the windward side. Newly planted trees will transpire more rapidly in windy situations, which can lead to the death of a tree already severely stressed by drought. The wind stability of trees is determined by tree species, stand structure, spacing, thinning regimes, soil classes, breeding and tree age at the time of anchorage.

The presence of airborne pollutants in the atmosphere has been a characteristic feature of the urban environment since the beginning of the Industrial Revolution. Air pollution can occur in a variety of forms but the principal ones are dust, SO_2 and NO_x. Leaves are the plant parts most likely to show symptoms of air pollution injury. On broadleaved plants, leaves may develop interveinal necrotic areas, marginal or tip necrosis, stippling of the upper surface, or silvering of the lower surface.

However, trees in the urban environment also play a role in the quest for cleaner air in the cities. Scott *et al.* (1998) showed that daily uptake of NO_2 and particulate matter represented 1-2% of anthropogenic emissions for the county of Sacramento, California.

In areas with winter temperatures below 0°C, the use of de-icing salt is a well-known problem. De-icing salt is applied to the surrounding environment by surface run-off, wet spraying and airborne drifting.

The initial and most common symptom of de-icing salt damage on trees and shrubs is reduced growth. This is often difficult to recognise or may be confused with other stress factors. Reduced growth is usually followed by early autumn colours and premature leaf fall. De-icing salt is usually accumulated on the windward side of trees. The damage is easily recognised because it faces the road and is normally regarded as the best indication of de-icing spray damage. The majority of trees and shrubs subjected to either soil salt or salt spray typically show necroses at the edges of the leaves or needles. Wounds, often related to pruning, are a common place for spray salt to infect the plants. The damage may cause lack of sprouting and eventually dieback. Conifers are very susceptible to de-icing salt spray damage because they are green all through the winter maintenance season. Trees and shrubs damaged by de-icing salt and showing dieback are difficult to cure.

Characteristics and Restriction of Rooting in the Built Environment

Urban soils as a growing medium are poorly understood and often misunderstood. Therefore plantings are carried out with little appreciation or attention to the character and quality of the material that lies beneath the surface. One major problem in relation to planting in the urban situation is soil compaction, which may occur in small as well as large urban sites. Soil compaction can be divided into two types: (i) intentional soil compaction, which occurs when soil is deliberately compacted for site stabilisation under roads, houses, etc. and (ii) unintentional soil compaction, which occurs when traffic uses areas intended for planting (Randrup 1997). In the urban situation, unintentional soil compaction is primarily found along roadsides and on construction sites.

When soil is compacted, its bulk density increases and its porosity decreases. These effects inhibit plant growth because the soil becomes impenetrable to root growth and, furthermore, restricts the water and oxygen available to the roots. For example, root growth of most plants is impeded once soil bulk density rises above 1.6. One consequence of compacted soil is waterlogging, which can kill roots around existing trees. Soil loosening has proved to be effective in alleviating compacted soil.

However, there is no doubt that the best treatment for compacted soil is to protect the soil from being compacted in the first place. Florgård (1987) suggested protecting trees growing on construction sites by dividing the site into zones in which different types of construction traffic are permitted. The principle of construction site zoning was adapted by Randrup and Dralle (1997) to protect the soil from being compacted. They suggested that the entire construction site be divided into a building zone, a working zone and a protection zone.

URBAN CHARACTERISTICS/ENVIRONMENT ANALYSED

The city is characterised by paving and buildings, which variously results in decreased wind speed, but an increase in windthrow, raised temperatures and precipitation, lowered humidity and shading in street canyons. The extent of the influences of these factors depends on the structure and amount of vegetation, as well as on the size of the city. Green areas in the city accessible to the public include areas at schools, public libraries and social institutions, as well as parks and churchyards. These are areas that are typically relatively small and geographically widespread, which is why they often suffer great recreational pressure, most often due to the overall proportion of green areas in the city.

Throughout Europe the proportion of urban green areas varies greatly, from over 60% of the area of Bratislava, the capital of Slovakia, to about 5% in Madrid, the capital of Spain. In comparison, the figure for Mexico City is only 2.2%; in relation to the number of inhabitants, this only provides 1.94 m^2 per inhabitant, which is far below the 9 m^2 recommended by the World Health Organisation. A suggested measure of urban environmental quality is the location of green areas within a walking distance of 15 min or less from all housing areas. This criterion is met for all citizens in Brussels, Copenhagen, Glasgow, Gothenburg, Madrid, Milan and Paris and, in general, for more than 50% of the population in most European cities.

The quality of urban green areas are increasingly recognised as being important to the overall quality of human life in the cities. Beyond this, the urban forests and trees are important as ecosystems in relation to the conservation of biological diversity.

Though the urban population benefits from the urban green areas, the increase in population places great pressure on the existing green areas as a result of urban and infrastructural development. The importance of the urban green space, of which urban forestry is an integral part, increases as population increases.

The growing urban population needs the environmental and social benefits associated with urban forests, as shown by Ulrich (1984) and Grahn (1989). In developing countries, urbanisation has had a dramatic influence on creating environments practically without any amenities, as is the case in Mexico City, where the growth in population has not been matched by an increase in green space. It is an accepted reality that the growth of the cities cannot be stopped. Instead, the challenge is to control urban growth so that it results in economic growth and a satisfactory environment.

As more than two-thirds of the population of Europe live in urban areas, the quality of the urban environment, including green areas, is becoming increasingly recognised as one key to the economic reconstruction of European cities (Commission of the European Community 1990). Urban areas constitute the everyday environment of the greater part of the population and, in recent years, this topic has received considerable attention in the European Union, United Nations and the Organisation for Economic Cooperation and Development, which will hopefully lead to an improvement in our knowledge of urban environments.

4

Climate Change in Forest

CLIMATE CHANGE IMPACTS ON FOREST HEALTH

The world's climate is changing. Increased temperatures and levels of atmospheric carbon dioxide as well as changes in precipitation and in the frequency and severity of extreme climatic events are just some of the changes occurring. These changes are having notable impacts on the world's forests and the forest sector through longer growing seasons, expansion of insect species ranges, and increased frequency of forest fires. Climate change can affect forest pests and the damage they cause by: directly impacting their development, survival, reproduction and spread; altering host defences and susceptibility; and indirectly impacting the relationships between pests, their environment and other species such as natural enemies, competitors and mutualists.A deeper understanding of the complex relationships between a changing climate, forests and forest pests is vital to enable those in forest health protection and management to expect and prepare for changes in pest behaviours, outbreaks and invasions. This chapter investigates these relationships and the implications for forest health protection and management. Specific examples where insect and pathogen lifecycles or habits have been altered by local, national or regional climatic changes are discussed. For example, the mountain pine beetle (*Dendroctonus ponderosae*) has shown decreases in generation time and winter mortality resulting in exponential population growth and major range extension in western North America. Warmer temperatures have resulted in range expansions of pests such as the pine processionary caterpillar (*Thaumetopoea pityocampa*) and the oak processionary caterpillar (*T. processionea*) in Europe and the southern pine beetle (*D. frontalis*) and red band needle blight (*Mycosphaerella pini*) in the US. Similar patterns are expected for other pathogens such as *Armillaria mellea*, and *Phytophthora cinnamomi* in Europe.

In the Democratic Peoples Republic of Korea, decreased winter temperature and lighter snow cover have contributed to reduced mortality and early emergence of overwintering *Dendrolimus spectabilis* (pine moth) with resulting loss of native *Pinus densiflora.* The purpose of this report is to compile and summarize the present knowledge on the impacts of climate change on forests and forest pests. Since limited research has been dedicated to determining such impacts on forest pests in particular, information on non-forest pests is provided throughout the report to enable a better understanding of the potential impacts of climate change on forest health.

CURRENT KNOWLEDGE AND FUTURE EXPECTATIONS

The world's climate is changing. While there are natural climate variations, the climate change we are most concerned about is in regards to the modifications to the greenhouse effect that human activities have caused. The Intergovernmental Panel on Climate Change (IPCC), in its Fourth Assessment Report, concluded with more certainty that global climate change is unequivocal and it is widely believed to result primarily from the effects of emissions of carbon dioxide (CO_2) and other greenhouse gases (GHGs) such as methane (CH_4) and nitrous oxide (N_2O), from human activities.

This conclusion was based on a number of observations about the Earth's climate including the following.

- Global surface temperature has increased by an estimated 0.74 degrees Celsius (°C) over the past century. Over a 50 year period from 1956 to 2005 the warming trend was nearly twice that for the 100 years from 1906 to 2005. Eleven of the 12 years from 1995 to 2006 rank among the 12 hottest years on record (since 1850, when sufficient worldwide temperature measurements began). Increases are widespread globally and are greater at higher northern latitudes. Over the last 50 years, cold days and nights as well as frosts have become less frequent over most land areas, while hot days and nights and heat waves have become more frequent.
- Consistent with warming, the extent of snow and ice has decreased. Mountain glaciers and snow cover have also declined on average worldwide. The maximum area of seasonally frozen ground in the Northern Hemisphere has decreased by about 7 percent since 1900, with decreases in the spring of up to 15 percent. Satellite data since 1978 show that the extent of Arctic sea ice during the summer has shrunk by more than 20 percent.
- Since 1961, the world's oceans have been absorbing more than 80 percent of the heat added to the climate, causing ocean water to expand thereby contributing to rising sea levels. This expansion was the largest contributor to sea level rise between 1993 and 2003. Melting glaciers and losses from the Greenland and Antarctic ice sheets have also contributed to recent sea level rise. The incidence of extreme high sea level has increased at a number of sites globally since 1975.

- From 1900 to 2005, significant increases in precipitation have been observed in eastern parts of North and South America, northern Europe and northern and central Asia. The frequency of heavy precipitation events has also increased over most areas. In contrast, precipitation has declined in the Sahel, the Mediterranean, southern Africa and parts of southern Asia. Droughts have become longer and more intense worldwide and have affected larger areas since the 1970s, particularly in the tropics and subtropics.
- Evidence of an increase in intense tropical cyclone activity in the North Atlantic has been observed since about 1970 and there are suggestions of similar increased activity in some other regions.

Predictions for the Future

The IPCC has made a number of predictions for future climate change. A warming of about 0.2°C per decade is projected for the next two decades; temperature projections beyond this period depend on specific emissions scenarios. Even if the concentrations of all GHGs and aerosols were kept constant at year 2000 levels, a further warming of about 0.1°C per decade would be expected.

The full range of projected temperature increase, based on six emission scenarios, is 1.1-6.4 °C by the end of the century. The best estimate range of projected temperature increase, which extends from the midpoint of the lowest emission scenario to the midpoint of the highest, is 1.8-4.0 °C by the end of the century. It is very likely that continued GHG emissions at or above current rates would cause further warming and induce many changes in the global climate system during the 21st century that would be larger than those observed during the 20th century.

Geographically the predicted warming trends for the 21st century are believed to be similar to those observed over the past several decades whereby temperature increases are expected to be greatest over land and at most high northern latitudes, and least over the Southern Ocean (near Antarctica) and the northern North Atlantic.

It is very likely that extreme heat, heat waves and heavy precipitation events will become more frequent. Increases in high latitude precipitation are very likely, while decreases are likely in most subtropical regions such as Egypt. It is also likely that future tropical cyclones (typhoons and hurricanes) will become more intense, with higher peak wind speeds and heavier precipitation associated with warmer tropical seas. Snow cover area is projected to contract, widespread increases in thaw depth are projected over most permafrost regions and sea ice is projected to shrink in both the Arctic and Antarctic. In some projections, Arctic late-summer sea ice disappears almost entirely by the latter part of the 21st century. Climate change is impacting the world's ecosystems and it is expected that the magnitude of these impacts will increase along with temperatures over this century. Many species and ecosystems may not be able to adapt as the effects of global warming and associated disturbances, such as floods, drought, wildfire and insect outbreaks, are compounded by other stresses such as land use change, overexploitation of resources, pollution and fragmentation of natural systems. If the global average temperature increases more than 1.5-2.5 °C, it is believed likely that approximately 20-30 percent of plant and animal species assessed so far will be at an increased risk of extinction. Major changes in ecosystem structure and function, species' ecological interactions and species' geographical ranges, with predominantly negative consequences for biodiversity and ecosystem goods and services are also projected.

FORESTS' EFFECT ON THE CLIMATE

Forests play an important role in regulating the earth's temperature and weather patterns by storing large quantities of carbon and water. This regulatory function has a profound effect on both the local and the global climate.

Locally, trees provide shade, which in turn lowers summer temperatures and prevents the soil from drying out, they reduce heat loss from the ground in winter and reduce storm damage by providing shelter from wind. Globally, forests regulate the global carbon cycle, having a profound effect on the climate. As well as this, deforestation is also contributing to climate change. Indeed the CO_2 released each year from forest loss is higher than that released by our yearly transport emissions.The continued existence of forests is particularly necessary if we are to halt what is known as runaway climate change. Runaway climate change is the point whereby increases in temperature lead to more GHG emissions which in turn leads to increased temperatures. Examples of this include that increased temperatures will melt ice caps which will lead to huge increases in the release of GHGs; and increases in temperature are projected to negatively affect up to two-thirds of existing forests, thereby exacerbating deforestation and increasing the release of carbon.

THE CLIMATE'S EFFECT ON FORESTS

Global warming, which on a geological timescale is occurring in the equivalent of a split second, is significantly disrupting the intricate and poorly understood web of interactions that governs the very structure and composition of forest ecosystems. This means that around a third of today's forests are likely to change their species composition. A temperature increase of 3°C by 2100 would result in forest ecosystems having to move 500 km towards the poles or 500m in elevation in order to find the same climatic conditions. Such distances are far beyond the average rate of dispersal for individual tree species, let alone entire forest ecosystems.

Early warnings about the consequences of the impacts of climate change on forests have been documented in, among others, The Carbon Bomb: Climate change and the fate of the northern Boreal forests - a 1994 Greenpeace report which states on page 2 that:

- "Studies on the global carbon cycle suggest that boreal forests are not absorbing as much carbon as they did before 1976. As a result, the atmosphere already appears to contain 10-15 billion tonnes of carbon more than it would have if forests had continued to absorb carbon at the pre-1976 rate. If boreal forests continue to decline, estimates suggest that burning and rotting of boreal forests could contribute to the release of up to 225 billion tones of extra carbon into the atmosphere, increasing current levels by a third. This would accelerate the rate of climate change."

While it is possible that the boreal forest could expand into the frozen tundra as temperatures increase, such an expansion would likely be delayed by slow tree migration rates. Even in the long-term, the boreal forest is unlikely to move northward fast enough to compensate for the breakdown of boreal forests at the southern part, turning dense forest into open woodlands and grassland, which in turn will result in a lowered biological diversity and a reduced ability of these ecosystems to store carbon and water.

Other forest ecosystems are faced with a similar fate; according to the Intergovernmental Panel on Climate Change(IPCC), a UN panel of climate scientists, it is likely that many tree species will not be able to change their geographic distribution fast enough to keep up with projected shifts in suitable climate and increased rates of extinctions are expected to occur.

What can be Done?

So trees (or the lack of them) are one of the key problems that must be tackled if we are to minimise the effects of climate change. It is not surprising then that many have also suggested that planting trees (carbon sinks) or reducing emissions from deforestation and degradation (REDD) are among the solutions to our climate crisis. SinksWatch believes however that while urgent action is needed to halt forest loss and restored degraded forests, both of these concepts - REDD and carbon sinks - are a dangerous distraction from tackling climate change.

Instead of addressing the root causes of the forest as well as the climate crisis, REDD and carbon sinks are concepts that are linked to a global carbon market where planting trees or reducing deforestation in one place (usually the global South) justifies even more fossil fuel emissions somewhere else (usually in an industrialised country). And many of these carbon sink projects have financed the expansion of large-scale industrial tree plantations. SinksWatch believes that efforts to tackle forest loss must not justify more fossil fuel emissions because such a trade-off would neither help avert runaway climate change not help save forests in the long run because runaway climate change will also have a major negative impact on the world's forests.

CLIMATE CHANGE COULD AFFECT ENTIRE FOREST ECOSYSTEMS

Off the southern coast of California, the fog rolls in over Santa Cruz Island coating the needles of the Bishop pines with moisture. Water drops form on the needles and roll down to fall to the forest floor, happening over and over on every needle of every tree. The forest ecosystem depends on this fog drip to stay alive. Bishop pines are thought to have proliferated all along the West Coast of the US and Mexico thousands of years ago when the environment was cooler and wetter. Today, the pines are found only on Santa Cruz Island, Santa Rosa Island, an island off Baja California, and in northern California where low-altitude clouds known locally as "June gloom," along with the cooler and wetter climate of the north help keep the trees growing.

A research team from the University of California — Santa Barbara, led by Mariah S. Carbone, studied the influence of clouds on the largest Bishop pine forest of Santa Cruz Island. "When people think about climate change, they're often thinking about temperature and precipitation," said Carbone. "When you think about precipitation, it's rain and snow, depending on where you are. What this study showed is that you can have really important water inputs coming from clouds that influence the carbon cycle."

Clouds are one of the largest uncertainties in global climate, according to Carbone, and the rather large impact they have on forest ecosystems has rarely been studied before, particularly in Southern California. The types of forest ecosystems that grow in coastal California will be altered by changes in cloudiness or cloud height that occur with global warming. While predictions about how California rainfall will change over the next half century are uncertain, scientists believe fog and low-stratus clouds might decline as sea-surface temperatures warm. Forests such as the relic Bishop pines on Santa Cruz Island and the Sequoias of coastal Northern California remove and store carbon from the atmosphere, making them important in understanding concentrations of greenhouse gases in the future. Additionally, these forests are prized for their aesthetic and recreational qualities.

From May through August, fog is often present in coastal California occurring when the land warms in the spring and summer as moist ocean air is pulled over the cold, upwelling coastal waters. This creates clouds or fog banks as the moist air condenses under a stable atmospheric inversion layer. "The finding that summer fog strongly impacts carbon cycling highlights the need for improved understanding of whether we should expect coastal summer cloud behaviour to change in a warmer world," said A. Park Williams, a former graduate student in UCSB's Geography Department, now at Los Alamos National Laboratory. "A change in summer fogginess could produce temperature, moisture, and carbon feedbacks in coastal ecosystems that easily swamp out the effects expected from increased greenhouse gases alone."

Billions of tons of carbon move between the oceans, lands and atmosphere every year in the carbon cycle, causing increasing amounts of carbon dioxide to be released into the atmosphere. This potent greenhouse gas contributes heavily to climate change. The UCSB team received help collecting information from both NASA and NOAA. "We used satellite data to map the spatial distribution of cloud cover across the western portion of Santa Cruz Island," said Christopher J. Still, formerly an associate professor at UC Santa Barbara, now at Oregon State University. "We had some evidence of a cloud cover gradient from the coast to inland, based on our own observations and our weather station data, but the satellite data —— from NOAA and NASA satellites —— really allowed us to quantify this gradient across the western half of the island and in particular at our two study sites." This study, linking clouds to soil microbial activity, is unique.

"While most previous research on fog and ecosystems has focused on its role in plant-water relations —— such as the well-known linkages between fog and coast Redwoods —— we show that soil microbial activity and metabolism in these coastal ecosystems may be very dependent on the light but frequent fog drip that occurs during the rainless summer months," said Still. Another unique aspect of the study is the linking of fog and low clouds to the carbon cycle of a coastal ecosystem and how this link impacts tree growth and soil respiration. "What we found was that there are two major effects of the clouds," said Carbone. "One is shading, which keeps the temperatures cooler and the soil moisture higher. The other is fog drip, which is a water input into the soil."

Where the trees were under more cloud cover, the soil moisture content depletes at a slower rate, allowing the trees to grow longer into the dry season. While water pulses from fog drip immediately stimulates microbes in the litter and soil beneath, it does not directly enhance pine tree activity. The microbial activity does, however, release carbon dioxide into the atmosphere.

"This study provides us with a greater understanding of how fog and the low coastal clouds which Californians commonly call 'the marine layer' affect plant growth and the activities of microorganisms in the soil," said John Randall, scientist with The Nature Conservancy. "Their findings of the importance of fog and low clouds in influencing plant growth and survival and microbial metabolism in the soil will help us understand how changes in coastal fogs and low clouds that may accompany climate change will affect the forests and soils on Santa Cruz Island, other islands off the coast of California and Mexico, and all along the coast of the adjacent mainland."

The Nature Conservancy owns and operates Santa Cruz Island in cooperation with the National Park Service and the University of California. From varied landforms supporting more than 600 plant species, to a wide diversity of animal life, to remnants of Native American and European settlements, Santa Cruz Island is an environment rich in biodiversity and cultural resources.

CAPACITY OF FORESTS TO RESPOND TO CLIMATE CHANGE

The adaptive capacity of forests, for the purpose of this document, is understood to be the inherent ability of the forest to adjust to changing conditions, moderating harms and taking advantage of opportunities. Thus, strengthening of the adaptive capacity is oriented at increasing the resistance or resilience to changes but may also include adapting the forest to new conditions by facilitating changes in the system (*e.g.* by species introduction). In general, strengthening the adaptive capacity of forests aims to maintain, restore or enhance forest area, biodiversity and forest health and vitality. Many of the actions oriented towards mitigation of climate change through REDD+ have a strong potential for synergies with actions oriented at strengthening the adaptive capacity of forests, in particular if such actions consider ecological safeguards, such as biodiversity conservation. Experiences in strengthening the adaptive capacity of forests to climate change have been more widespread in plantations and agroforestry systems.

These systems tend to have a simpler structure and composition that makes it easier to detect changes due to climate change and to design and implement adaptation-strengthening mechanisms. This is much more difficult in complex natural forests, in particular in the tropics. However, because of their simplicity, these systems may also be more vulnerable and therefore the need to look at adaptation options is greater. Interestingly, several of these adaptation activities are oriented towards making these (agro) ecosystems more diverse.

Maintaining Forest Area

Larger forests usually have greater species diversity and cover a greater variety of sites, thus reducing the risk of losing the whole system if climate change negatively affects several species or specific site conditions. Forest management appears to be the solution; both wellmanaged protected areas and well-managed community and private forest concessions in Guatemala and the South of Mexico have shown to be more effective in avoiding deforestation, fires and forest degradation than areas poorly managed areas. This probably goes beyond purely economic considerations.

Land and forest tenure, recognising the benefit of maintaining forests and joining forces with other forest managers are all important requisites. Size of forest area also seems to be important, for both individual landowners and community or multiple owners. Ecologically, larger forest areas show less edge effects, while from the management point of view, larger sizes allow for economies of scale.

Managing natural forests often is recognized as a claim on that forest. If good relations are held with local people, such claims are well respected. Owning forest but not managing it, has often resulted in unauthorized entry by third parties for the extraction of timber and NWFPs, or for conversion to agricultural land. Managing the forest but not entertaining good relations with the neighbours has often resulted in forest use conflicts, at times ending up in armed conflicts or burning of parts of the forest estate.

Such relations are more important in large forest tracks, since in these it is harder to establish continuous human occupancy. Good forest management normally includes fire, pest and disease management. Of these, fire management may be the most significant in maintaining the forest area, although serious pests, such as the mountain pine beetle in pine forests in North America, may also contribute to substantial forest loss. Managing the forest may be costly and income from the sale of one or more of its products may not off-set the extra cost of management. However, often, costbenefit analyses compare conventional operations (without much strategic planning or considerations for biodiversity or forest dependent communities) with managed operations.

From a private forest owner's point of view, this may be reasonable, but in practice, this approach has been used to justify continued conventional harvesting operations, giving the forest sector a poor image and increasing the pressure on governments to impose stricter regulations. In countries with greater willingness of the private sector to participate in 15 improving forest management, a series of alternatives were found to either make forest management attractive or propose other forest-based income solutions. Usually this was done by a carrot and stick approach: if a forest manager does better than the legislation requires, they receive subsidies, discounts on taxes and are a preferred provider of specific ecosystem services.

While these approaches in theory seem to be very promising, in practice they have not had the expected outcomes. This is partially because of the high financial and administrative cost to actually obtain the carrots and because forest owners and managers were not aware of the existing opportunities.

Management of forests, therefore, should go beyond the mere planning of protective or productive activities and not depend on one single form of financial income. Forest managers need to be informed and aware of local, national and international opportunities for income generation. New opportunities may be through tax or fee discounts for good forest management (as the case of harvesting fee discounts for certified forests in Peru), payment for environmental services (*e.g.* Costa Rica and Mexico) or niche markets (*e.g.*

markets for specific NWFPs and carbon). Due to the relatively limited demand of these above mentioned products and services and with only a few exceptions (*e.g.* Brazil nut gathering in Bolivia, Stoian, 2004), none of these have been able to single-handedly pay for management and protection of large natural forest tracts. Only where combinations of products and services were obtained has management of natural forests become a serious land use competitor. New opportunities may also work for forest plantations, where payment for environmental services, such as carbon sequestration, may at least partially off-set the initial establishment costs. However, selling sequestered carbon at the end of the rotation has little effect on the overall profitability of plantations, due to the low carbon price and high interest rates.

The present value of these future sales is very low and rarely will change a non-profitable exercise into a profitable one. Selling less carbon earlier on during the rotation, or shortening rotation length, are two options that may make timber plantations more attractive. Establishing tree plantations not for timber, but only for carbon or other environmental services, as yet has to show its profitability for the forest owner or manager and is usually only accomplished where the forest owner or manager also garners important non-tangible benefits from those forest.

Conserving Biodiversity

Maintaining forest area is of course a good means to maintain a certain level of biodiversity. However, as can be seen from some countries, increasing forest area does not necessarily increase biodiversity nor does it necessarily mean that old forests or undisturbed forests are maintained. In many countries, reduction of net deforestation figures is at least partially due to compensatory measures, such as natural regeneration and plantations. Although depending on how these new forests are being managed, and how close they are (in time and space) to the original natural forests, these usually do not have the same species composition and biodiversity as the lost natural forests.

In terms of capacity to adapt to climate change, the change in species composition may sometimes be an advantage, if new species are better adapted to changing conditions. More problematic may be loss of diversity. Loss of diversity will make forests more vulnerable to changes, since they will not have the rich gene and species pool from which to select for the new conditions. In this respect, care should be taken of the trade-offs between mitigation and adaptation objectives; too great an emphasis on management for carbon may reduce structural and compositional diversity, thus reducing the system's inherent adaptive capacity.

A noted change in forest management in the light of climate change has therefore been an increased interest in maintaining or increasing diversity of the forests. Mixed species plantations, use of a larger number of clones and reductions in the scale of harvesting operations have been implemented as measures to maintain or increase biological diversity. These same measures are now receiving more attention because of their potential benefit in preparing forests for climate change. In addition, literature suggests that the potential yield increase from 16 appropriately selected species mixes will more than outweigh the additional costs that may be involved in mixed tree plantation establishment. The use of nitrogen fixing tree species as part of the mix, in particular in degraded lands, is beneficial for overall growth rates.Reducing the scale of harvesting operations is one way of increasing the possibility of ecological connectivity between forest patches. Plantation establishment is also an important measure that may achieve this as does planting of trees outside the forest.

Although some practices are being adopted and theoretically will contribute to maintaining biodiversity, there is still a need for further research. For example, we do not yet know how much biodiversity change will cause a major and irreversible change of forest types or even ecosystems. Some authors suggest that maintaining functional diversity may be sufficient. However, it is unclear how susceptible diversity will be to climate change, if, for the different functions, the number of species that provide diversity function is strongly reduced. It is also unclear how species will react to climate change, by themselves, in combination with other species and in combination with a number of environmental and human factors. Continuous monitoring of the forests is critical to providing insight into these interactions.

Maintaining Forest Health and Vitality

The main threats to health and vitality are pests, diseases, fires and extreme weather events. In addition, diversity usually strengthens health and vitality and therefore the actions mentioned above to maintain or enhance biodiversity, also are useful for maintaining health and vitality. A number of silvicultural techniques have been developed for maintaining health and vigour of a stand. Removing old, poorly formed and damaged trees, for example, reduces the risk of spreading diseases and pests, although at the same time it may reduce diversity and thus increase the susceptibility of forests to diseases and pests. Applying such treatments requires knowledge of the specific risks related to individual tree species and the potential benefits of maintaining poorly formed trees in the forest. Using harvesting residues on the forest floor may increase availability of nutrients for the remaining trees, thus increasing vigour, but may add to the fuel load and therefore increase fire risk. Nevertheless, timing of use (beginning of wet season) may increase the benefits and reduce the risks.

In plantation forests, a reduction of old growth and an increase in the relative presence of young stands, enhances the general health and vigour of the forest from the point of view of timber production in the medium term.

Again, however, it reduces diversity, thereby reducing the adaptive capacity of the forest to externally driven changes. The decision to reduce such growth in favour of young stands needs to be taken in consideration with local conditions and management objectives.

Reducing Risk and Intensity of Damage

Reducing the risk and intensity of pests, diseases, fires and hurricane damage, along with managing the hydrological cycle, will become major concerns for many, if not all, forest managers under changing climatic conditions. Due to the complexity of measures that may have contradictory effects, there is the tendency for integrated management practices; for example, combining insect control with monitoring exercises and implementation of management practices that reduce susceptibility of the forest to insect attacks. Such practices include those treatments that help maintain the vitality of the forest, including timely thinning and species mix. For most regions however, few comprehensive management plans exist, and in most cases, plans emphasize monitoring and combating pests and diseases, rather than preventing them.

The example of pine beetle management in Central America shows how management oriented towards the prevention of fires may also reduce the risk of insect pests. Integrated fire management in forests has been promoted in many countries and by international and national agencies alike.

The proceedings of a 2009 seminar in California summarize many of the experiences and achievements to date. Martell (2007) defines forest fire management as: "getting the right amount of fire to the right place at the right time at the right cost". Integrated fire management includes the following components; prediction of fire occurrence, fire prevention, fire detection, initial attack, fire management, strategic planning of resources, and fuel management. Training, knowledge sharing and planning of the different components of fire management is extremely important.

Prediction of fire occurrence requires the identification of the main causes for forest fires, followed by an analysis of factors that influence the frequency of those causes. In Mexico, this has been done through proxy indicators, such as nearness of populations and types of land use (danger assessment). This needs to be combined with characteristics of the fuel, temperatures and moisture content (risk assessment) as well as with an assessment of the value of the resource in order to be able to set priorities in fire prevention and combat activities. In many developed countries, this information is combined with daily weather forecasts to assess fire danger ratings. This has proved to be a useful early warning system.

Fire prevention is probably the most cost effective and efficient way to reduce fire damage. It requires identification of frequency, size, severity and causes of human induced fires and needs to be followed by awareness campaigns to reduce those causes or projects oriented at training and introduction of alternatives to burning forest and grazing lands. In many regions, for example Mediterranean countries and western states of the United States, watch towers are the main tool for fire detection. Patrol flights are used as well but it is difficult to plan them cost effectively. Once detected communication of location, size and burning characteristics of the fires among the fire fighters and to the public is very important, especially for the initial attack.

Whereas in North America it is a problem of deciding how many air tankers to keep on standby for the initial response to fires, in many tropical countries air tankers are not available and initial attack may need to be done by local fire brigades. The quicker the response time, the smaller will be the risk of uncontrolled fires.

However, such fire brigades require an extensive programme of training of the local teams, as well as sufficient equipment to be able to fight small fires. Initial attacks become more efficient if these are based on previously prepared plans that consider potential impact of fire fighting, as well as the potential risks of damage to natural, human and physical resources.

Fire management requires continuous monitoring of fuel conditions, fire behaviour and climate. With this information and short-term predictions on their changes, fire fighters can decide when and how many people to dispatch to control the fire. Fuel management is one of the more common and discussed approaches to forest fire reduction. Protection of forests has led in many cases to the accumulation of fuel, increasing the risk of forest loss, rather than reducing it. Prescribed burning is a common tool, but this needs to be supervised by experts, in order to avoid the fires to get out of control. Cutting of fire breaks is another measure often taken.

These do not only form barriers to fire progress, but may also help accessibility of the forest for fire fighting crews. Technology has proved vital in monitoring and control of forest fires in regions such as North America and Europe. However, in tropical countries, where access to this technology and to the forests is more limited, fire management strategies need to include all stakeholders and all related sectors since most fires originate outside forest lands and involve non-forest stakeholders.

Risk reduction strategies for damage from hurricanes or cyclones need to consider damage to forest, agricultural crops, communities and infrastructure. Reducing risk of damage to forests can be done through silvicultural practices such as shorter rotation cycles; young trees often are more resistant to wind throw, but if thrown, will result in less biomass lost compared to larger trees.

Increasing resilience, for example by maintaining good seed sources and mixed species forests, including species that readily sprout after wind throw is another means of reducing damage. In Toncontín, Honduras, the local community started to improve forest management in 1997, one year before hurricane Mitch struck. The forest is dominated by species resilient to hurricanes through different strategies including rapid and abundant seed regeneration or re-sprouting. Part of the management improvement was the retention of seed trees. Comparative studies one year after Hurricane Mitch indicated that the forest with seed tree retention was recovering its original species composition faster than forests where all commercial trees had been cut without seed tree retention.

FORESTS OF THE FUTURE BECOME CARBON SOURCES

Just to put the terrestrial biosphere in perspective, there is about ten times more carbon contained in all land plants (plus the soil they grow on) than all the "extra" anthropogenic carbon currently in the atmosphere. Most of the terrestrial carbon is contained in forests which have been significantly depleted by mismanagement. The question is whether Northwest forests are more likely to store or release carbon under a changing climate. The coupled processes of photosynthesis and respiration/decomposition mirror each other at a global scale to help regulate CO_2 levels and our climate. Photosynthesis captures water and CO_2 and liberates oxygen to create biomass, while respiration consumes biomass and oxygen to liberate CO_2 and water.

Depending on temperature and moisture conditions, among other factors, photosynthesis sometimes dominates leading to net carbon uptake. At other times respiration/decomposition dominate leading to net carbon release. Whether our forests ultimately become net carbon sources or net carbon sinks under the future climate of the Northwest depends on factors that remain uncertain, such as the amount of summer precipitation vs. drought stress, the effects of future climate on fuels and fire hazard, the effects of CO_2 enrichment and climate change on plant physiology, whether forests geographically expand or contract, and whether forests are exploited or protected.

The good news is that slight to moderate warming may increase our forests' ability to store carbon through increased growth and geographic expansion. Pacific Northwest forests might become significant carbon sinks and help mitigate climate change if growing conditions remain favourable and disturbances like fire do not significantly increase. Under warm-wet conditions growing seasons will lengthen, and forest or woodland communities could expand into current rangelands, thus raising the possibility that northwest forests could absorb CO_2 and become a significant net carbon sink.

The bad news is that there is likely a warming *threshold* above which our forests will likely decline due to drought stress and increased disturbances. Drought stress limits the potential photosynthetic benefits of longer growing seasons and CO_2 enrichment. Increasing temperature also increases rates of respiration and decomposition, so under a future climate scenario like this, northwest forests could wither, recede geographically, and become a significant net carbon source. The IPCC tells us that some warming has already occurred and that existing levels of CO_2 already commit us to some additional warming. There is considerable uncertainty about when we may cross the sink to source threshold.

El Niño/Southern Oscillation (ENSO) is a prominent source of multi-year variability in weather and climate around the world. The main signature of ENSO is a periodic (~every 3-8 years) reduction in winds moving westward across the Pacific ocean. This allows warm water to move eastward across the tropical Pacific Ocean. ENSO has strong impacts on ocean nutrient cycling and associated fish populations and birds. ENSO has repercussions far beyond the Pacific ocean, including periodic wide-scale drought in many regions of the world.

Scientists have found a correlation between periodic phenomena like ENSO and years with anomalous global increases in CO_2 which appear to be linked to CO_2 releases from plants, soil, and fire. While there remains debate about this, some have predicted that ENSO may become more frequent and sustained under global warming which could cause a positive feedback favouring respiration over photosynthesis on a global scale.

The source/sink differences could also manifest themselves differently across geography and time periods. "In regions where drought stress is not important because of high levels of precipitation, or if increases in CO_2 concentration increase water use efficiency and thus reduce water stress, longer growing seasons could result in increased growth. Where drought stress is important, a longer growing season may mean only that plant respiration exceeds photosynthesis for a longer time, which would result in reduced growth." So, it is conceivable that moist forests west of the Cascades might remain net carbon sinks, while the dryer forests east of the Cascades might become net sources.

Another study looked at the effects of CO_2 enrichment and climate change on vegetation in the mid- and high-latitudes of the northern hemisphere and found opposing effects in spring and summer. CO_2 uptake was apparently enhanced during warm wet spring season, but looking over the entire growing season, including the dryer summer, CO_2 uptake did not increase. Another paper estimated that western forests might increase in spatial extent while decreasing in their carbon density, *i.e.*, more forested acres, but fewer trees per acre.

CAPACITIES OF FOREST MANAGERS TO RESPOND TO CLIMATE CHANGE

Climate change is already affecting forest ecosystems and will have increasing effects on forests and their capacity to supply of goods and services. These impacts of climate change will have far-reaching social and economic consequences, particularly for forest dependent communities and the forest dependent poor. The challenge of climate change for forest managers was recently recognised in the recent report of the IUFRO Global Forest Expert Panel on Adaptation of Forests to Climate Change.

Many forest managers are aware of climate change and that it may have positive or negative effects on their forests, but only few have a good understanding of the degree of uncertainty of climate change and impact projections. They are thus susceptible to wrongly interpret those projections unless they receive capacity building on forests and climate change. Most of the existing and proposed actions for mitigation or adaptation to climate change do not require different technical capacities to those already required for SFM. However, it is often not clear to managers that existing best practices may be adequate for the expected or occurring climate induced changes in the natural environment. Even without considering climate change, many forest managers are not up to date with the best practices for SFM. Preparing for climate change, therefore, needs to include capacity building in SFM practices, including integrated fire management, integrated pest and disease management, harvest planning (including road building), silvicultural practices and the implications actions for nutrient and water cycles. The practices to be focused upon during capacity building will depend on the country and target group. In the Mediterranean, for example, training for fire management will be more relevant whereas in tropical American pine forest, pests and diseases may be a more relevant topic, although fire management is also important in these areas. In addition, there is a general need to strengthen the monitoring capacities of forest managers. Awareness raising on adequate monitoring methodologies, tools, and indicators and threshold values that will allow for early detection of changes is also required. This is relevant especially for managers focussing on timber production, forest conservation and protection. Although the livelihoods of forest dependent people are largely based on NWFPs, there is relatively little awareness of the ecological requirements and management practices of these NWFPs. Capturing that knowledge and sharing it with others will greatly contribute to strengthening adaptive capacity of many communities.

Appropriate Technology

Appropriate technology is needed for three major lines of action: monitoring and research, adaptation and mitigation and will vary according to geographic area, type and objective of forest management activity, scale and intensity of operation and existing local human and financial resources. The major problem with the availability of technology is not its existence, but its location and adaptiveness to local conditions. It may be as simple as having instructions for the use and maintenance of equipment only in a foreign language, not readily understood by the intended users. The cost for acquisition and implementation of this technology is another common problem.

Monitoring and Research

Monitoring and research of vegetation cover usually requires a range of instruments, from simple measuring tapes to highly sophisticated remote sensors. In addition it requires specific software that allows storage and analysis of the data. While instruments and software exist, they are often not accessible for, or need to be adjusted to the requirements of forest managers. The experience with software for forest inventories shows that tailor-made software has a tendency to get out of date rapidly and after a few years can no longer be used within the new operational systems. It is therefore becoming more common to look for commercial software. The large Canadian based forest products company Tembec, for example, has used a package of commercially available software as basis for their ISO 14001 monitoring system. This, however, requires qualified personnel to combine the software and adjust it to the company's needs. Small and medium scale forest managers, especially in developing countries, will need to collaborate with research and monitoring organisations to be able to make the necessary adjustments in a cost effective way. Adaptation may require new irrigation technology, as well as breeding technology and seed storage technology. Mitigation, on the other hand, will need technology efficient in energy use.

GAPS IN THE INSTITUTIONAL ENVIRONMENT

For the purpose of this document, the institutional environment is considered the framework of formal and informal arrangements that will set the limits for the implementation of climate change preparedness activities. This environment includes the normative framework (legislation and policies), markets and other arrangements to finance forest management and conservation, as well as informal arrangements that direct behaviour of the local actors.

Property Rights

Ownership and property rights are a major concern in many regions of the world, particularly for indigenous people in developing countries. There are some cases where rights and tenure are protected by legislation but enforcement is oftentimes lacking. In Costa Rica, for example, land, tree and carbon rights are well-defined, but still a considerable part of the defined rights in the northern zone of Costa Rica are currently under dispute. In other countries, poor people are characterised by not having legal access to land.

Normative Framework

While many countries have recently revised or developed forest legislation, the normative framework for forests and climate change needs to go well beyond legislation and policies of the forest sector. Commerce, transport, agriculture and finance are just a few of the other sectors that have a great influence on what happens in forestry. Mechanisms to foster this intersectoral collaboration are often nonexistent in most countries and often climate change is dealt with outside of the forestry sector.

The different actors of the forestry sector are often not well-organized at the local, national or international levels and have little experience in transparent policy setting and implementation. In many cases this has led to a lack of trust, which makes collaboration even more difficult.

Financial Arrangements

Investment costs of change are high, while at the same time forest owners or users do not have access to financial mechanisms that help cover those costs. REDD+ talks have created great expectations amongst developing countries however there have been many obstacles thus far for the successful implementation of REDD+ activities. It is thought that the costs for REDD+ implementation may exceed the capacity of developed countries. Further, developing countries do have the absorption capacity for the REDD+ funds. This may leading to uncontrolled spending and may do more harm than good in the long term. Markets for carbon from trees and forests are still not well established and there are no generally acceptable and workable definitions of permanence, additionality and leakage or measuring and monitoring systems in place. Few effective PES schemes exist, mainly because the institutional structure is not in place or too expensive to operate.

ADEQUATE MANAGEMENT RESPONSES TO CLIMATE CHANGE

The enabling conditions for the forest management responses to climate change are identified and the challenges that these gaps may create analysed. It should be noted that relatively few experiences in forest management responses to climate change have been documented and therefore it is difficult to provide a comprehensive overview of existing capacities and those needed for adaptation and mitigation actions. Lack of knowledge on climate change impacts on forests In many tropical countries, information on forest resources (type, quality, extent, values, and changes) is deficient. This information can come from studies on specific topics, (*e.g.* tree growth of individual species under various conditions), or from local or national monitoring, gathering data with a predetermined frequency on a predetermined set of variables relevant for forest management and conservation. For the more general information, such as forest area, forest quality and species composition, monitoring is the preferred method. In Europe, the United States and some tropical countries, such as Mexico and Chile, such information comes from continuous national forest inventories. In most tropical countries, however, such monitoring mechanisms are limited and sometimes non-existent.

NATIONAL MONITORING SCHEMES

In a revision of national monitoring schemes related to regional SFM standards (such as the Tarapoto process, the Montreal process and the ITTO criteria and indicators for SFM) at a workshop in Chile in April 2011 (Günther *et al.*, in prep.), many similarities were identified between the standards at the criterion level, but countries found it difficult to harmonize monitoring at the indicator level. This was partially due to differences in priorities, but largely also due to differences in capacities to measure the indicators and in the institutional arrangements necessary to measure and share the results between the different state and private entities involved.

This is in particular the case for socio-economic data, but also for the quality of forest resources. The quality of monitoring, furthermore, depends on continuity of the measurements, which requires long term funding (often not available in tropical countries), clear and constant definitions of the variables and processes to be measured (such as forest degradation and deforestation and forest quality) as well as compatible measurements over time (also not available for many indicators). The FAO Forest Resource Assessment (FRA) process was recognised as providing a good base to set up and harmonize monitoring tools and mechanisms. However, it was noted that this process suffers from the same obstacles as the existing national and regional processes.

At the project or community scale, monitoring often encounters even greater obstacles, due to the disadvantages of measuring at smaller scales. While it may be possible to simplify monitoring mechanisms, this is only possible if the monitoring results are adequate for the scale, intensity and risk level of the proposed forest operations. Even in certified operations, where monitoring at project scale has developed further than in most other operations, only few well designed and operational monitoring systems exist that actually contribute to improved forest management.

While such monitoring is essential for good forest management at any scale, it needs to be accompanied by monitoring of weather conditions in order to be a useful tool for forest managers to prepare themselves for climate change. In many countries there is no detailed information on weather conditions at the local scale, limiting the ability to relate past and current forest behaviour to weather conditions. In addition, the lack of such data reduces the capability to downscale climate change projections and thus the ability to project climate change impacts at a local scale. It is therefore necessary to increase the number of weather stations throughout much of the tropics (especially the large forest areas), as well as the capacity to analyse the data and incorporate them into impact and climate change models at a local scale.

Research

Research is usually used to solve specific problems or establish links between actions and impacts. It may also be used for a more in-depth understanding of forest processes and the contribution and influence of human activities on forests. Research on climate change impacts and adaptation options have been focused mainly on boreal and temperate forests. Little information exists on impacts of climate change on tropical forests and how tropical forests will react. While potential changes can be studied in field and laboratory trials, it will also be necessary to improve climate change projections at a small scale. Only with reliable local climate change projections will forest managers be able to plan for future changes. Data on the response of species and individual trees to changing environmental conditions is particularly important for plantations, where a limited number of species is planted and thus substitution may not take place when conditions change. For the major timber species, more work is required to identify their levels of resilience to climate change. This information will also be relevant for mitigation, since tree dynamics and its change under changing climatic conditions will influence the capacity to sequester carbon. Research on functional groups may be as relevant specifically for adaptation of forests. There is a lack of case studies highlighting successful cases of forest managers implementing adaptation strategies. Currently, most studies look at tree or animal species. Few have a system approach that includes the environmental, economic as well as the social and technical aspects of forest management.

Much is already known about the relationship between trees and water. However, this relation may vary according to geographic, topographic, geological and soil conditions. More studies are required to understand local influences of trees and forests on the hydrological cycle, especially in those areas where climate change is likely to change the amount and the distribution of rainfall over time. Climate induced changes in the water cycle (*e.g.* increased frequency and severity of droughts and floods), may affect the capacity of local forests and people to adapt to climate change as well as the mitigation potential of forests. Further, there are several myths surrounding the beneficial effects of trees and the negative effects they may have on water availability, soil erosion and other potential ecosystem services. Undoing those myths may be harder than learning from new experiences.

The social and cultural impacts of climate change on forests dependent communities have not been studied widely. Changing forests may change the availability or quality of ecosystem services, resulting in changes in the relationship between people and forests, social relations among people and the degree to which cultural needs are met. This could lead to migration to other areas in search of the lost ecosystem services or other livelihood opportunities. One of the consequences of migration is increased conflicts over resources. There is an urgent need to study these potential conflicts and to recommend options to manage these conflicts. REDD+ is being widely studied, particularly at the project and national levels. The impact that REDD+ strategies may have on individual households in the medium and long term needs to be further studied to understand the potential contribution that REDD+ may have to local development and biodiversity conservation. While this may not be of direct concern to forest managers, it will, in the long term, be important to avoid social conflicts and the results of such studies may alter the REDD+ mechanisms that are in the process of being designed.

Communication

Monitoring and research will be of little use if the results are not communicated to potential users and the general public at large. IUFRO, among others, have published guidelines for effective communication in the forestry sector and there are many cases where these guidelines have been successfully applied. In general, however, communication within the sector is weak, resulting in many stakeholders not having access to reliable and updated information. They may miss new opportunities or even act upon incorrect information.

5

Forest Plantations and Biodiversity a Fresh Perspective

Forest plantations covered 187 million hectares in 2000, of which Asia accounted for 62 per cent. The forest plantation area represents a significant increase from the 1995 estimate of 124 million hectares. The reported new annual planting rate is 4.5 million hectares globally, with Asia and South America accounting for 89 per cent. About 3 million hectares are estimated to be successful. Globally, half the forest plantation estate is for industrial end-use, one-quarter for non-industrial end-use and one-quarter not specified. Globally, the main fast-growing, short-rotation species are in the genera *Eucalyptus* and *Acacia*. Pines and other coniferous species are the main medium-rotation utility species, primarily in the temperate and boreal zones.

The potential for forest plantations to partially meet demand from natural forests for wood and fibre for industrial uses is increasing. Although accounting for only 5 per cent of global forest cover, forest plantations were estimated in the year 2000 to supply about 35 per cent of global roundwood. This figure is anticipated to increase to 44 per cent by 2020. In some countries forest plantation production already contributes the majority of industrial wood supply. There is increasing interest in development of forest plantations as carbon sinks; however, failure to resolve international debates on legal instruments, mechanisms and monitoring remains a serious constraint.

In developing countries about one-third of the total plantation estate was primarily grown for woodfuel in 1995 although it should be noted that planted trees on farmland, in villages and homesteads and along roads and waterways contribute significantly to fuelwood supplies, enabling the demand to be met in most instances.

New forest plantation areas were reported as being established globally at the rate of 4.5 million hectares per year, with Asia and South America accounting for more new plantations than the other regions. Of plantations established, about 3 million hectares per year were estimated as being successful. Of the estimated 187 million hectares of plantations worldwide in 2000, Asia had by far the largest area. In terms of composition, *Pinus* spp. (20 per cent) and *Eucalyptus* spp. (10 per cent) remain dominant worldwide, although the overall diversity of species planted was shown to be increasing. Industrial plantations account for 48 per cent, non-industrial plantations for 26 per cent and plantations for unspecified use for 26 per cent of the global forest plantation estate.

The results of the plantation assessment were the first global estimates with a uniform definition of forest plantations and can therefore not be directly compared to previous estimates. FRA 2000 country statistics on plantations may also differ from those reported in prior FAO publications, partly because of changes in definitions. Countries participated directly in the assessment, providing technical documentation and supporting analysis and validating the results generated by FAO. Several experts around the world were enlisted to provide detailed information on various aspects of the plantation situation in the form of special studies.

Between the extremes of afforestation and unaided natural regeneration of natural forests, there is a range of forest conditions in which human interventions occur. European forests have long traditions of human intervention in site preparation, tree establishment, silviculture and protection; yet these are not always defined as forest plantations. The traditional forest plantation concept tends to be applied to single species, uniform planting densities and even age classes. Terms such as "natural forest under management" or "assisted natural regeneration" are applied to stands of indigenous species in more heterogeneous management mechanisms in Europe and other industrialized temperate and boreal countries.

METHODS

The area of existing forest plantations would ideally all have been derived from statistically designed inventories of forest plantations or statistics for planted areas reported by planting agencies or appearing in national reports. However, information also comes from many other sources including nursery production, seedling distribution and estimates derived from the goals of planting programmes.

The vast range of agencies, industries and non-governmental organizations within countries engaged in planting programmes made the comprehensive collection of all relevant source documents a major logistical exercise. For FRA 2000, over 800 source documents were analysed to derive the forest plantation estimates. In most developing countries a national clearinghouse for collecting information on plantations is either lacking or ineffective owing to the enormity of the task and limited resources.

Data Collection

To retrieve the source documents for the plantation study, FAO made formal requests to all developing countries, some of which contributed the necessary materials. Most of the reports were collected directly by FAO staff during FRA 2000 workshops and visits to national ministries. For consistency FRA 2000 prepared guidelines and questionnaires for the collection of forest plantation statistics in which the objectives, scope, definitions, sources of data and templates for specific data collection were supplied to each country. Parameters requested included:

- Total estimated forest plantation area, 2000;
- Annual area of new plantations;
- Species groups: broadleaf, conifer, non-forest like, African oil palm (*Elaeis guineensis*), coconut palm (*Cocos nucifera*) bamboo or unspecified;
- Purpose and end-use objective of forest plantations: industrial (producing wood or fibre for supply to wood processing industry) or non-industrial (fuelwood, soil and water protection);
- Ownership: public, private, other (*e.g.* traditional, customary) or unspecified.

Other data requested in the guidelines, which proved difficult for countries to provide by species group, included age class distribution; end-use by forest product (industrial plantations); growth and yield (mean annual increment); standing volumes; and rotation lengths. Despite the absence of these data, FRA 2000 is the most comprehensive forest plantation resources assessment that has been carried out.

In previous assessments of forest plantation resources, plantation data were available up to the reference year for most countries, since the reporting followed the reference year. In FRA 2000, the reference year was 2000, so if data were not available to that date, then existing area and annual planting data were used to extrapolate the necessary information. For the few countries that have no data sets since 1990, the rate of planting in preceding years and future planting programmes were considered in projections to the year 2000.

FAO also enlisted the assistance of several experts around the world to make specific technical contributions on the forest plantation situation in the 1990s. These studies constituted an important part of the global results as well and complemented the country information.

Analysis and Interpretation

The quantity and quality of forest plantation data provided is dependent upon the capacity of the national forest inventory systems to collect and analyse data and to adjust the information to conform with global and regional reporting parameters. In many developing countries there is a lack of institutional capacity to carry out periodic national forest inventories, so data can be incomplete, inconsistent, outdated and of variable reliability. Because of this, it was necessary to derive and in some instances to verify forest plantation statistics through desk research using available country reports. All sources of country data were referenced and made available in a transparent manner.

In addition, regional and national focal persons were appointed to assist in the forest plantation data collection, to ensure that the latest data were available and to maintain coordination and communication between FRA 2000, FAO regional offices and each participating country. On completion of the data sets, a formal verification process was undertaken with each participating country.

Purpose and ownership within the global forest plantation estate Purpose and ownership of forest plantations vary markedly among regions. Industrial plantations provide the raw material for wood processing for commercial purposes, including timber for construction, panel products and furniture and pulpwood for paper. In contrast, non-industrial plantations are aimed for example at supplying fuelwood, providing soil and water conservation, wind protection, biological diversity conservation and other non-commercial purposes.

In many countries, particularly in the developing world, the end purpose of the plantations is not clearly defined at the outset. In some of these cases, valuable tree resources are established which coincidentally match future needs. However, in other cases the lack of planning may result in plantations that have little commercial value and a low potential for local use.

SELECTED GLOBAL TRENDS, 1980-2000

Comparisons

FRA 2000 country statistics on plantations may differ from those reported in prior FAO publications partly because of changes in definitions. For example, rubber plantations were not previously considered as forest plantations but are included in FRA 2000 plantation data. Previous assessments also used regional reduction factors to indicate the successful proportion of plantations remaining after establishment. The FRA 2000 assessment applied reduction factors according to the best available data from each country independently. There have also been changes in the information base from which the estimates were derived. The statistics now include data from many industrialized countries, none of which were included in the prior global assessment reports. Despite these differences, comparison of FRA results from each decade allows analysis of some trends including planting rates, genera, areas and purpose.

IMPACTS OF THE FOREST PLANTATION ESTATE

The potential for forest plantations to partially meet demand for wood and fibre for industrial uses is increasing. According to FRA 2000, the global forest plantation area accounts for only 5 per cent of global forest cover and the industrial forest plantation estate for less than 3 per cent. However, as an indication only, forest plantations were estimated in the year 2000 to supply about 35 per cent of global roundwood and an increase to 44 per cent anticipated by 2020. If plantation development is targeted at the most appropriate ecological zones and if sustainable forest management principles are applied, forest plantations can provide a critical substitute for natural forest raw material supply. In several countries industrial wood production from forest plantations has significantly substituted for wood supply from natural forest resources. Forest plantations in New Zealand met 99 per cent of the country's needs for industrial roundwood in 1997; the corresponding figure in Chile was 84 per cent, Brazil 62 per cent and Zambia and Zimbabwe 50 per cent each. This substitution by forest plantations may help reduce logging pressure on natural forests in areas in which unsustainable harvesting of wood is a major cause of forest degradation and where logging roads facilitate access that may lead to deforestation.

Forest plantations also provide additional non-wood forest products, from the trees planted or from other elements of the ecosystem that they help to create. They contribute environmental, social and economic benefits. Forest plantations are used in combating desertification, absorbing carbon to offset carbon emissions, protecting soil and water, rehabilitating lands exhausted from other land uses, providing rural employment and, if planned effectively, diversifying the rural landscape and maintaining biodiversity.

Not all forest plantation development has positive economic, environmental, social or cultural impacts. Without adequate planning and without appropriate management, forest plantations may be grown in the wrong sites, with the wrong species/provenances, by the wrong growers, for the wrong reasons.

Examples exist where natural forests have been cleared to establish forest plantation development or where customary owners of traditional lands may have been alienated from their sources of food, medicine and livelihoods. In some instances poor site/species matching and inadequate silviculture have resulted in poor growth, hygiene, volume yields and economic returns. In other instances, changes in soil and water status have caused problems for local communities. Land use conflicts can occur between forest plantation development and other sectors, particularly the agricultural sector.

The negative impacts of forest plantations can draw the focus away from the fact that forest plantation resources are totally renewable and can be economically, socially, culturally and environmentally sustainable with prudent planning, management, utilization and marketing.

SELECTED FOREST PLANTATION TOPICS

Mean Annual Volume Increment (MAI) of Select Industrial Species

On average *Eucalyptus* and *Pinus* species, which dominate industrial plantations in developing countries, have similar MAIs of 10 to 20m^3 per hectare per year. However, many of the popular species of both genera frequently achieve much faster growth rates. Thus *Eucalyptus grandis,* which is the most widely planted *Eucalyptus* species, can achieve 40 to 50 m^3 per hectare per year and in very exceptional conditions with advanced tree improvement 100m^3 per hectare per year. Other widely planted tropical hardwoods including *Casuarina equisetifolia, Casuarina junghuhniana*, *Tectona grandis* and *Dalbergia sissoo* have MAIs of less than 15m^3 per hectare per year and frequently under 10 m^3 per hectare per year.

Climate and site have a very large impact on growth rates. The humid tropics and more fertile sites are more conducive to higher growth rates than locations with long dry seasons or infertile or degraded soil. Teak on many sites in India, for example, frequently has an MAI of 4 to 8 m^3 per hectare per year, partly because of drought combined with poor soils. Some species such as *Gmelina arborea* and some of the *Eucalyptus* species are very site sensitive. *Pinus* spp., in contrast, generally tolerate adverse conditions better and are more flexible with respect to site. Both tree breeding and silviculture have improved growth rates. Good examples are *Eucalyptus grandis* and *E. urophylla* in Brazil and *Pinus radiata* in some countries of the Southern Hemisphere.

Advanced silviculture typically includes improved nursery and establishment techniques such as good site preparation, weed control and judicious use of fertilizer. It has been suggested that growth of teak, for example, could be doubled in Kerala, India and Bangladesh and increased sixfold in Indonesia by adopting these practices.

With coppice species productivity varies with rotation, the first and second coppice rotations usually being more productive than the seedling one. The growth patterns vary among species. For example, very fast growing species such as *Gmelina arborea* can reach a peak MAI in less than 10 years, while *Pinus caribaea* var. *hondurensis* grown in Trinidad reaches maximum MAI at about 25 years and *P. radiata* at over 40 years. With *Cupressus lusitanica* in Costa Rica, the MAI maximum is reached at about 30 years.

Rotation lengths can reflect both end-use and economics. Many fast-growing *Eucalyptus, Acacia* and *Casuarina* species and *Gmelina arborea* are grown on short rotations of under 15 years as they are used primarily for pulp or woodfuel. Usual rotations in Kenya for *E. grandis* are 6 years for domestic woodfuel, 7 to 8 years for telephone poles and 10 to 12 years for industrial woodfuel. In Brazil this species is largely grown for pulp or charcoal on 5 to 10 year rotations. Species being grown for high-value sawlogs usually have longer rotations; teak (*Tectona grandis*) is grown on 50 to 70 year rotations and high-value conifers such as *Araucaria angustifolia* on 40 year rotations. Generally pines are grown on medium-length rotations of 20 to 30 years, unless grown solely for pulpwood, when shorter rotations may be adopted.

Modelling of growth, rotations, harvest yields and product mix by species is important for decision-making in forest management. One of the major obstacles to model development for planners and managers is a lack of suitable data. Data can come from a range of sources, including temporary and permanent sample plots and experiments. Experiments and protocols for obtaining data need to be carefully designed so that reliable information is obtained over the complete range of conditions to which the model is to apply. Tree Growth and Permanent Plot Information System (TROPIS), sponsored by the Centre for International Forestry Research (CIFOR), seeks to coordinate and improve access to tree growth information.

A growth and yield model developed for *Pinus elliottii* plantations in coastal Zululand, South Africa, predicts height, basal area, total stem and merchantable volume and stocking, with age on a stand level for harvest planning. In New Zealand several simulation models have been developed for *P. radiata* which predict similar variables but also include wood quality, harvesting and marketing aspects, which make it possible to link the silvicultural options to industrial use.

Sustaining Productivity

It is possible not only to sustain but also to increase productivity in successive rotations. This requires clear definition of the end-use objective for forest plantation development and a holistic view in their management. There is a need to integrate strategies for genetic improvement programmes, nursery practices, site and species/provenance matching, appropriate silviculture (site preparation, establishment, weeding, fertilizing, pruning, thinning), forest protection and harvesting practices with prudent management. New Zealand and the southern United States have shown that substantial gains can be made by adopting this holistic approach.

In developing countries where resources may be constrained, highly technical solutions may not be essential but it is critical to get the fundamentals correct: careful species and provenance choice, good nursery stock, site preparation, planting techniques, weed control and, less frequently, fertilizer inputs. Once fast-growing, uniform plantations have been established, later silvicultural tending may become increasingly important, depending on the end-use objective.

Current evidence suggests that plantation production can be sustainable if foresters implement prudent genetic and silvicultural tree improvement programmes and sound management practices. There has been, however, limited long-term research on the subject; there are few definitive studies, limited to few species.

In one of the most promising studies, with *Pinus patula* in Swaziland grown intensively on about 15 year rotations, site productivity was maintained or increased over three rotations. The question of declining growth in teak (*Tectona grandis*) plantations in Indonesia and India remains unclear.How forest plantations are managed affects the chemical and physical properties of the soils and site. However, only recently have long-term studies been undertaken to evaluate these critical factors or processes.

The methods adopted for site preparation (ripping, ploughing, scarifying, bedding, windrowing, controlled burning), establishment (manual, mechanical), weeding (manual, chemical, mechanical), fertilizer application, pruning and thinning (manual and mechanical, for commercial to waste), forest protection and harvesting (manual, mechanical, clear-fell or selection) all affect the pool of nutrients in the ecosystem. Interference with the drainage, litter and recycling of organic matter and change in the physical conditions of soils during these operations are critical to long-term sustainability.

Because of litter recycling and the rapid development of tree roots, plantations are used for rehabilitation of fragile and degraded lands prone to soil erosion and excessive water Run-off.

Tree plantations often have higher evapotranspiration rates than grassland or agricultural crops and thus change the hydrology of the site. This can be either beneficial (for example by reducing salinity problems in some dryland conditions) or detrimental (if it reduces water required for other uses).

The rare studies of changes of productivity between rotations have concluded that negative changes have primarily been due to inappropriate or inadequate management practices or weed invasions rather than a result of the plantations themselves.

Burning and excessive cultivation in site preparation, soil compaction from mechanical operations, inappropriate harvesting techniques and poor forest protection can contribute to loss of nutrients and soil erosion, with a resultant loss in productivity of forest plantation sites. This cannot be addressed solely by addition of fertilizer, but by the adoption of the whole range of tree improvement, silviculture, protection and harvesting techniques in an integrated forest management strategy.

Valuable Hardwood Plantations

Long-rotation, slow-growing but valuable hardwood species have special technical properties, such as strength, natural durability, hardness and easy machining and appearance (grain, figure, texture, colour and other aesthetic qualities) that make them suitable for high-value end-uses such as furniture.

These high-grade hardwoods contrast with short-rotation, fast-growing, lesser-quality woods used for woodfuel, pulpwood or reconstituted products and less demanding building timbers. In tropical countries teak (*Tectona grandis*), mahogany (*Swietenia* spp.) and rosewood (*Dalbergia* spp.) are the main hardwood plantation species, while in temperate countries oak (*Quercus* spp.), ash (*Fraxinus* spp.), cherry (*Prunus* spp.), walnut (*Juglans* spp.), tulipwood (*Jacaranda* spp.) and hard maple (*Acer* spp.) predominate.

Table. Characteristics of Valuable Hardwoods used in Tropical Areas

Use categories	*Desirable wood properties*	*Main end-uses*	*Matching valuable hardwood species*	*Comments*
Decorative timbers	Appearance, consistent quality, dimensional stability, durability, good machining, staining and finishing properties	Quality furniture and interior joinery	*Tieghemella* spp.; *Entandophragma cylindricum*, *Chorophora* spp., *Aucoumea klaineana*, *Afrormosia* spp., *Entandophragma utile*, *Mansonia* spp., *Lovoa* spp., *Khaya* spp., *Swietenia* spp., *Dalbergia* spp., *Aningeria* spp.	Highest value, competitio n from temperate hardwoods and MDF
High to very high-density timbers	Appearance, strength, high natural durability, availability in large sizes	Principall y in constructi on	*Dipterocarpus* spp., *Lophira* spp., *Chlorophora* spp., *Ocotea rodiaei*	Small share of total tropical timber use
Low to medium-density utility timbers	Appearance, clear grain, natural durability, good machining properties	External joinery, shop fittings, medium-priced furniture	*Shorea* spp, *Hevea brasiliensis*, *Terminalia* spp., *Heritiera* spp.	Most commonly used, prone to competitio n from substitutes

Because many valuable hardwood species are difficult to establish because of their ecological requirements or disease or insect susceptibility, focus has been on the easier species to grow, including teak (*Tectona grandis*), Indian rosewood (*Dalbergia sissoo*) and mahogany.

In 1995 the global areas of these species were 2 254 000, 626 000 and 151 000 ha, respectively.

They accounted for about 10 per cent of total hardwood plantations in the tropics. More than 90 per cent of teak plantations were located in Asia, mainly in Indonesia, India, Thailand, Bangladesh, Myanmar and Sri Lanka. About 95 per cent of rosewood plantations are located in India and Pakistan. The largest mahogany plantations are located in Indonesia and Fiji, which together make up about 80 per cent of the established area. The market preference for large piece sizes, slow growth and very long rotation lengths combine to reduce the attractiveness for commercial investment in these species.

This is only partially counteracted by their value. The low return on capital investment, coupled with the long wait period for this return, has made it difficult to interest private investors without supportive, secure and stable government policies.

As markets demand a continuity of supply, plantations need to be on a sustainable scale within a region. Some of the less common species are not known in the marketplace. Other potential market problems are that the timber may be wrongly associated with tropical deforestation and changing fashions that often occur with decorative timbers. Niche marketing is important for valuable hardwoods.

Projections for supplies of timber from existing valuable hardwood plantations indicate that because of the age class distribution and long rotations there will not be a significant increase in supply in the next 20 years. Future promotion of quality hardwood plantations needs to emphasize choice of species with versatile end-uses, market research and development to hold on to niche markets and maintained high standards from production to marketing.

Careful site selection, use of high-quality planting materials of superior genetic origin and good silviculture are important. Planting programmes should be economically viable, environmentally appropriate and socially desirable. Incentives may also be necessary to stimulate private investment because of the long rotations.

Even though valuable hardwood plantations have the potential to reduce the pressure on natural forests, they will not prevent deforestation resulting from agricultural encroachment.

The supply of large quantities of high-value timber could perhaps undermine the value of natural forest stands and so lead to more rapid destruction. Hence it is advisable, where possible, to manage plantations and forest resources and forest products in a complementary manner.

Plantations and Wood Energy

Woodfuels from plantations or natural or semi-natural forests are particularly important in developing countries, providing about 15 per cent of their total energy demand. Woodfuel provides about 7 per cent of energy demand for the world as a whole and in industrialized countries only 2 per cent. Woodfuel provides more than 70 per cent of energy needs in 34 developing countries and more than 90 per cent in 13 countries. Woodfuel makes up about 80 per cent of total wood use in developing countries and about 89 per cent in Africa

The prediction of a woodfuel crisis in developing countries in the 1980s was based largely on looking at supply and demand from forest plantations and natural forests. The reaction to the expected woodfuel crisis was to plant trees for this purpose, often in the form of traditional plantations. Many programme failures resulted from lack of appreciation for the complexities of bioenergy supply and demand, failure to take into account social aspects and people's needs and poor programme structures. The importance of planted trees on farmland, in villages and homesteads and along roads and waterways as a source of woodfuel supply was underestimated.

Rural communities harvest stems, branches, stumps, twigs, leaves and litter for woodfuels in chronic woodfuel supply areas. In these instances the nutrient recycling process is broken, resulting in degradation of forest plantation sites. In many rural communities in developing countries, woodfuel is considered a public free good, to be foraged from public natural and plantation forests. Often women and children collect the woodfuel at little or no cost. As a result, the growing of private forest plantations specifically for woodfuel, in which development costs and rotation cycles are involved, can be a foreign concept.

Asian studies show that forest-based supply can range from 13 per cent in the Philippines to as high as 73 per cent in Nepal. In many countries less than 50 per cent of fuelwood is from forests.

Globally, non-industrial forest plantations in 1995 were estimated to cover about 20 million hectares. This was almost 17 per cent of the world's total plantation area in 1995. A significant proportion of these plantations were planted for woodfuel and 98 per cent were in developing countries. These plantation figures do not account for trees planted outside the forest on farms or in villages, etc., nor do they consider plantations that were considered agricultural plantations, such as *Hevea* or palm plantations.

In developing countries about one-third of the total plantation estate was grown primarily for woodfuel in 1995. Three-quarters of these plantations were in Asia, where they accounted for 60 per cent of total plantation production. In Latin America more than half of plantation production went to woodfuel; in Africa and Occania a larger proportion of plantation production was as industrial wood. However, plantations, in general, provided only a small proportion of total woodfuel used. Uruguay is an interesting exception.

Production of woodfuel from plantations currently makes only a small contribution to energy requirements, although it is very important in some localities and countries. Plantations currently supply 5 per cent of woodfuel. Production from these non-industrial plantations is likely to double over the next 20 years, even with little expansion in area, because the age class distribution is heavily concentrated in young plantations. In an optimistic scenario where planting continues at the same rate as in the past ten years and then gradually declines, a 350 per cent increase in woodfuel production would be anticipated by 2020. By-products from wood-using industries will also contribute to increased fuelwood supply. The situation is less positive in Africa, where for a few countries declines are projected in plantation-based woodfuel production.

New Sources of Fibre

Since FRA 1990, advances in wood utilization technology have resulted in increasing importance of new sources of fibre - rubber (*Hevea brasiliensis*), coconut palm (*Cocos nucifera*) and African oil palm (*Elaeis guineensis*) - especially in the Southeast Asian subregion.

These species account for 9.7, 12.0 and 6.0 million hectares of plantations, respectively. All grow in the humid tropics. In terms of plantation area, Asia has 92 per cent of the world's rubber, 86 per cent of the world's coconut palm and 78 per cent of the world's African oil palm. Indonesia,

Thailand and Malaysia have almost three-quarters of the rubber plantations; Indonesia and the Philippines have about half the coconut resources; and Malaysia has 55 per cent of the oil palm resource. All three species are grown principally for other products rather than wood, so when overmature they are available for fibre-based industries at minimal cost.

Rubberwood is harvested when latex productivity declines and yields 100 m^3 per hectare of roundwood, but recovery for lumber is only 25 to 45 per cent because of poor form and small size. Most of the planted stands in Southeast Asia are owned by smallholders and are geographically dispersed, with poor accessibility and poor-quality stems. Currently the major proportion of industrially utilized rubberwood comes from large-scale plantations. Quality furniture, parquet, panelling, reconstituted panels, general utility timber and woodfuel, including charcoal, are made from rubberwood.

However, the rubberwood must be processed within days of harvesting to minimize sapstain attack. The most developed downstream industries are in Malaysia, where the production of sawn rubberwood timber rose from 88 000 m^3 in 1990 to 137 000 m^3 in 1997 and medium density fibreboard (MDF) production from rubberwood reached 1.16 million cubic metres per annum by 1999. Exports of rubberwood furniture have grown from about US$74 million in 1991 to US$683 million in 1998. Rubberwood has become a substitute for light tropical forest hardwoods. Its acceptance as a sustainable plantation-grown, environmentally friendly timber has given it wide appeal.

Coconut palms are harvested as the copra yields decline (beyond 60 years) and yield 90 m^3 per hectare of coconut wood. Coconut palm has variable properties and is intrinsically difficult for conversion but can yield a relatively low-cost, general-utility timber for construction, panelling, stairs, door jambs, furniture, flooring and power poles.

In 1993 Indonesia had 65 million cubic metres of overmature coconut stems which needed disposal before replanting. There is increasing interest in this raw material in European and North American markets. It is unlikely to replace conventional timber, but likely to find its way into niche markets. It will continue to be used as a low-cost construction timber.

Oil palm plantations are harvested for fibre beyond the 25 to 30 year rotations and yield about 235 m^3 per hectare. It is estimated that over 1.6 billion cubic metres of fibre will be available in the years to come from established resources in Southeast Asia. From 1996 to 1999 the area increased by 18 per cent. In Malaysia the area has increased by 3 million hectares in the past 30 years.

Most oil palm plantations (unlike rubber and coconut) in the main growing countries, Malaysia and Indonesia, are managed by plantation companies or cooperatives. Oil palm by-products such as kernel shells, pressed fibres and empty fruit bunches are currently used in heat generation at the extraction plants. Water in the stems can reach five times the weight of dry matter.

The high moisture content as well as the high amounts of parenchyma tissue rich in sugar and starches make conversion into quality forest products a challenge. An MDF plant in Malaysia is currently being planned to utilize oil palm stems.

Plantation Substitutes for Natural Forest Products

With growing concerns about the status and loss of natural forests, the rapid expansion of protected areas and large areas of forest unavailable for wood supply, plantations are increasingly expected to provide substitutes for products from natural forests, particularly in Asia and the Pacific.

In Asia and the Pacific it is estimated that 52 per cent of natural forests are not available for wood harvest because they are inaccessible or uneconomic to exploit. Of the unavailable forest in the region, it is estimated that about 38 per cent is legally reserved. In addition, logging bans have been imposed on large areas of natural forest covering about 10 million hectares. The reasons for these bans vary but were related to deforestation and forest degradation causing environmental problems in Thailand, the Philippines and China and to conservation requirements in Sri Lanka and New Zealand.

As a result of the net effect of deforestation and removal of natural forests from wood production, some areas in the Asia and the Pacific region have wood deficits and roundwood harvesting is exceeding sustainable levels of cut. The worst affected areas are South Asia and insular Southeast Asia, with continental Southeast Asia also under strain. In contrast, New Zealand has surplus plantation wood available for export.

Of six examples studied in the Asia and the Pacific region, New Zealand is more than self-sufficient in wood production based on plantations. In China and Viet Nam, the importance of plantations will increase as planted resources mature. There have been serious problems with implementing plantation development programmes in Sri Lanka, the Philippines and Thailand. In Sri Lanka, India and elsewhere in the tropics, trees outside the forest are playing a critical role in roundwood and woodfuel supply.

Most countries in the region are becoming importers of wood, with imports expected to rise. Sometimes logging bans have shifted the problem to other countries. Problems with acquiring large areas of land in some countries make it difficult to implement industrial plantations. In the Philippines, Thailand and Viet Nam there have been social conflicts with local indigenous people or between traditional forest use and development, as well as between the rich and the poor. Sometimes incentives and the development of social forestry programmes are being used to help resolve such problems.

While it is clear that plantations will have an increasingly significant role in substituting products from natural forests, the impact will be felt on a case-by-case basis as governments and investors determine where and how plantations can be technically, economically and socially feasible as well as environmentally friendly. In the near term, plantations in Asia and the Pacific can make a contribution but cannot replace harvests from natural forests. It is likely that both in the region and globally the current pace of industrial plantation development will barely keep pace with losses from deforestation and transfer of natural forests to protected status. While it would be theoretically possible, actual plantation development is at present not sufficient to offset both growing consumption and declining harvest from natural forests.

Plantations and Carbon Sequestration

In the past ten years, the development of forest plantations as carbon offsets has evolved towards a market mechanism, although an organized market with carbon prices defined according to supply and demand forces is still a long way off. The adoption of the Kyoto Protocol in 1997 triggered a strong increase in investment in plantations as carbon sinks, although the legal and policy instruments and guidelines for management are still debated.

A number of countries have already prepared themselves for the additional funding for the establishment of human-made forests. The 1997 Costa Rica national programme was the first to establish tradeable securities of carbon sinks that could be used to offset emissions and the first to utilize independent certification insurance.

To date, greenhouse gas mitigation funding covers about 4 million hectares of forest plantations worldwide. The recognition of afforestation and reforestation as the only eligible land use, land use change and forestry activities under the Clean Development Mechanism of the Kyoto Protocol, as agreed in Bonn during the second part of the Sixth Conference of the Parties to UNFCCC in July 2001, will lead to a steep increase in forest plantation establishment in developing countries. The sink decision of the Bonn Agreement is expected to funnel additional funds into forest activities in developing countries and thus to strengthen the international efforts in this field. However, it will also require a monitoring and verification system to ensure that these plantations will not be established at the expense of the local population or efforts to conserve biological diversity. Thus the decisions taken in Bonn to make the Kyoto Protocol ratifiable will also bear new challenges for forest plantation development.

Trees Outside the Forest

The significance of trees outside the forest (TOF) can be observed in several contexts. In countries with low forest cover, TOF resources constitute the main source of wood and non-wood "forest" products, even though trees may be so scattered that the maps produced by FRA 2000 indicate that no forests exist.

Trees are found on agricultural lands, in densely populated areas, in fruit-tree plantations and in home gardens, which often cover a large proportion of the land. In urban areas trees provide important aesthetic and environmental services in addition to providing shade and greatly increasing the livability of cities. Communities, farmers and herders who do not have access to forests diversify their production and protect their land by maintaining various tree systems on their farms.

Deforestation has been mapped and quantified, but very little is known about the fate of land formerly under forest; forest clearing is often followed by the establishment of production systems of which trees are an integral part. Not much is known about the dynamics of trees on farmlands and their corresponding contribution to the production of wood and other products and services. Similarly, little is known about changes in tree cover in fields and urban systems. Knowledge of trees outside the forest comes mostly from local studies on agroforestry, sylvipastoralism and urban, social, community or rural forestry.

This widespread and multipurpose resource, familiar to farmers but poorly defined by managers and mostly absent from official statistics and development policies, needs to be better assessed and known.

Growing populations, shrinking forests and degraded ecosystems all suggest that trees outside the forest are destined to play a larger local and global role in meeting the challenges of resource sustainability, poverty reduction and food security. Trees outside the forest relieve the pressure on forest resources, conserve farmland, boost agricultural productivity, blunt the harmful impact of urban growth on the environment, increase food supplies, provide income and in general make valuable contributions to food security.

FRA 2000 did not undertake a global assessment of trees outside the forest, mainly because of resource limitations; nor has there ever been a comprehensive global assessment of trees outside the forest and their products. However, a number of studies have been carried out for specific sectors or geographic areas, often with an emphasis on their economic contributions. This chapter provides a summary of selected studies and discusses the practical and conceptual difficulties related to a comprehensive global assessment.

The chapter responds to the concern expressed by the Expert Consultation on Forest Resources Assessment 2000 regarding the lack of information on TOF.

6

Forest Conservation and Environments

Interactions between tree species and between them and disturbance regimes generate distinctive assemblages of species, forest dynamics and landscape patchworks. These are best illustrated by examples at the ends of the range of disturbance frequency.

HIGHLY DISTURBED ENVIRONMENTS

Where disturbances are frequent, large scale or severe, forests tend to be dominated by intolerant and fast-colonizing species. Three examples illustrate the range of circumstances.

Floodplain Forests

Floodplain forests are influenced by channel movement, which destroys mature stands but reworks the deposits into new shoals. In northern temperate regions, these are colonized mainly by *Salix, Alnus* and *Populus* species, which grow into even-aged, often monospecific stands. The sequence of channel movements is manifested as a pattern of elongated even-aged stands, whose age increases with distance from the channel. Tolerant species colonize beneath these pioneer stands and, given sufficient time, develop into mixed old-growth. In northern temperate deciduous regions the principal long-term dominants are *Ulmus, Fraxinus, Quercus, Carya, Sassafras* and, in the Pacific North-west, *Picea sitchensis* and *Pseudotsuga menziesii.* At any one time, the pioneer stands tend to predominate near the present channel, and the mixed old-growth tends to survive in elongated patches at some distance from the river.

Ice-dominated Forests

The classic type is the wave-regenerated *Abies* forests of the eastern USA and Japan. Exposed mature forest degenerates when foliage is stripped by ice and wind from trees that have already lost the vitality of youth. Death of exposed trees exposes others, leading to a 'wave' of mortality, which moves steadily through the forest in the direction of the prevailing wind. Regeneration starts within the degenerating stands and grows vigorously in the lee of slightly older stands. The forest as a whole takes the form of a series of parallel waves, which move through the forest at 1-3 m annually on a return time of 60-70 years. This perpetual recycling ensures that *Abies balsamea* remains dominant and that the longer-lived *Picea rubens* is perpetually excluded.

Fire-dominated Forests

The most widespread form of highly disturbed forest is dominated by fire. Most boreal forests are naturally fire-dominated, but so too are Mediterranean forests and the forests that fringe extensive grassland and desert regions. Each region has a suite of species that undergo a characteristic succession after fire, for example in Scandinavia *Betula-Pinus* develops into *Pinus* dominance, which is then succeeded by *Picea.*

Exceptionally, some patches remain unburned, which allows the pioneers to be completely displaced, although generally fire returns in good time to ensure that pioneer species remain a permanent feature of the forest. In fact, there is much variation in return time, associated with variation in topography, ground vegetation and the configuration of water bodies. Nevertheless, fires were frequent enough to maintain most boreal forests as young or maturing stands, not old-growth.

Similar fire-dominated regimes control other forests, such as the *Eucalyptus* forests of Australia and the *Pinus*-dominated forests of the coastal plain of the south-eastern USA. However, in western North America many tree species not only withstand fires once they have achieved a moderate size but also grow to great size and age, thus generating the monumental forests of *Sequoiadendron giganteum, Sequoia sempervirens* and *Pseudot-suga menziesii* in which old individuals may bear the scars of several fires. On the margins, the boundary between mesic- and fire-dominated forests advances and retreats according to the history of fires.

Relatively Undisturbed Environments

At the other end of the range are mesic forests growing in relatively undisturbed environments. These are not disturbance-free; instead, catastrophic disturbances are rare enough to allow most of the forest to develop into old-growth, where disturbances are small-scale events.

Over much of the north temperate zone, a distinction can be drawn between *Fagus* forests and mixed deciduous forests. *Fagus*-dominated forests are found mainly in northern latitudes and submontane elevations, where *Fagus* spp. often share dominance with conifers, for example *Tsuga canadensis* in the eastern USA, *Abies alba* and *Picea abies* in central Europe. European beech, *Fagus sylvatica,* is not long-lived, but casts dense shade, is capable of regenerating in the small transient gaps generated in beech forests and is almost fireproof, so that it is able to both dominate the site and perpetuate this dominance.

However, it is prone to disaster, in the sense that mature stands are vulnerable to drought and high winds. A few intolerant species, such as *Salix caprea,* maintain a foothold in large gaps, while the shade-tolerant *Acer pseudoplatanus,* like *Acer saccha-rum* in the eastern USA, fills gaps, competes in advance regeneration and can grow into mature stands. Diversity is maintained partly by the tendency of species not to regenerate under themselves. Thus, *Fagus grandifolia* and *Acer saccharum* in the USA and *Fagus syl-vatica* and *Abies alba* in Europe have been reported to alternate, thereby maintaining a small-scale mosaic of different dominance, though in the former case coexistence has been ascribed to different responses to light intensity. Occasionally, such stands are destroyed by storms, whereupon pioneer species dominate the regrowth, although beech thrives in the underwood and eventually restores its position.

Mixed deciduous forests are those which lack the dominating influence of beech. At their greatest development, for example in the southern Appalachians, they comprise a mixture of several dozen species, each with the capacity to occupy the canopy or subcanopy. A wide range of genera are represented, notably *Acer, Quercus, Fraxinus, Carya, Aescu-lus, Betula, Castanea, Tilia, Magnolia, Carpinus* and *Halesia,* forming rich mixtures in which no single species becomes absolutely dominant. Such forests are rarely devastated by any single disturbance and for most of the time are renewed by gap-phase regeneration on a small scale. Pioneer species regenerate in the larger gaps and may dominate the canopy after periods of enhanced gap creation. An example is *Lirioden-dron tulipifera,* which not only forms the tallest trees in the Appalachian forests but can also live for 500 years, enabling it to perpetuate itself through long periods lacking disturbance.

Within these complex mixtures, each species has a distinctive pattern of growth, longevity and regeneration. Furthermore, there is a tendency for individuals of one species to be replaced by another species. The forests comprise a small-scale mosaic of groups of different canopy and underwood species, each with its particular successional trend. The trends in one patch are countered by opposite trends in other patches, thereby retaining the mixture. Recruitment of particular species tends to be irregular, depending on particular combinations of mast-years and disturbances. Composition remains fairly constant overall but at a small scale changes perpetually, except where individual trees replace themselves vegetatively, *e.g. Tilia* spp.

Single-species groups are common in mixed forests. These may develop in response to small differences in site conditions and/or the chance coincidence of an episode of gap creation with heavy seed production by a particular species. Simulations have recently shown that neighbourhood effects may also play a part.

Where there is a high probability that canopy trees will be replaced by individuals of the same species, 10 generations is enough for this feedback to generate small-scale single-species patches. The scale of the patches increases substantially where minor environmental differences result in 5% alterations in recruitment probabilities.

FOREST BIODIVERSITY CONSERVATION IN ENVIRONMENT

In the Convention on Biological Diversity signed by many member states at the Earth Summit held in Rio de Janeiro (Brazil) in 1992, explains biodiversity as follows: "Biological diversity" means the variability among living organisms from all sources including terrestrial, marine and other aquatic ecosystems and the ecological complexes of which they are part; this includes diversity within species, between species and of ecosystems.

TYPES OF BIODIVERSITY

Biodiversity is a generic term that can be related to many environments and species, for example, forests, freshwater, marine and temperate environments, the soil, crop plants, domestic animals, wild species and micro-organisms. Basically it can be classified according to three types of diversity:

- ecosystems and landscapes (habitat diversity)
- animal, plant, bacterial species (species diversity)
- all genes (genetic diversity)

Of particular importance are the taxonomically isolated species, as they have little similarity to other species and therefore are unique with respect to their genetic structure. These species are often endemic meaning limited to one specific area. Their extinction would mean a greater loss for global biodiversity rather than just the extinction of a species.

Distribution of Biodiversity

In Europe, biodiversity is very unevenly distributed with the least variety of ecosystems and lowest diversity occurring in Northern Europe. Centres of high biodiversity are to be found in the Mediterranean (Italy, Spain, Greece, France) and on the fringes of Europe (Bulgaria, Ukraine, Georgia, Armenia, Turkey), with over 5,000 endemic plant species that occur only in these countries. The Mediterranean is Europe's richest sea in terms of biodiversity.

The Importance of Biodiversity

Biodiversity is often used to draw attention to issues related to the environment. It can be closely related to

- the health of ecosystems.

For example, the loss of just one species can have different effects ranging from the disappearance of the species to complete collapse of the ecosystem itself. This is due to every species having a certain role within an ecosystem and being interlinked with other species.

- the health of mankind.

Experiencing nature is of great importance to humans and teaches us different values. It is good to take a walk in the forest, to smell flowers and breath fresh air. More specifically, natural food and medicine can be linked to biodiversity.

Biodiversity in the Black Sea

The Black Sea region used to be one of the most important areas for fisheries and for food and income for local people. Sturgeons (Acipenser sp.), mullets (Mugil sp.), and mackerel (Scomber sp.) as well as other species have been extensively exploited in this area. However, human activities related to agriculture, shipping and tourism now exert strong pressure on the environment, especially in the northern part of the Black Sea, and are taking their toll on biodiversity and damaging fish stocks.

Threats to Biodiversity

Large amounts of fertilizers were carried to the sea from agricultural areas along the large rivers, such as Don, Dnepr, and Dnjestr. Oil leaked into the sea from ports on the eastern shores and directly from tankers crossing the Mediterranean. Wastewater from cities and heavily visited coastal tourist towns was discharged without any purification.

Consequences for the Ecosystem

All of this leads to pollution and eutrophication meaning the enhanced blooming of planktonic algae due to increased nutrient loads. Planktonic algal blooms lower the water transparency letting only a little amount of light through the water where makrophytes usually grow. This is how the natural belt of bottom vegetation along the Black Sea coast has been destroyed. For example, the vertical distribution range of Cystoseira spp. decreased from 0-10m to 0-2,5m. The consequences were drastic because the habitat is of major importance as a nursery for spawn and hatchlings of many marine species. This led to a drop in reproduction rates and hence fish stocks.

Biodiversity of the Coastal Zone

Coastal areas are exceptionally productive environments, rich in natural resources, biological diversity and with a high potential for commercial activity. The importance of biodiversity in the coastal zone can be demonstrated by 8 out of the 40 EU listed priority habitats of wild fauna and flora falling into the coastal habitat. Approximately a third of the EU's wetlands are located on the coast as well as more than 30% of the Special Protected Areas designated under the Directive for the conservation of wild birds. The reproduction and nursery grounds of most fish and shellfish species of economic value also comes from this area, which accounts for almost half of the jobs in the fisheries sector.

Pressure on Coastal Biodiversity

Coastal areas are increasingly vulnerable to stresses from both human activities and the forces of nature. The complexity of human activities, natural systems and ownership in the coastal zone, requires an integrated management scheme to allocate coastal resources efficiently and minimize environmental degradation. Choices have to be made between competing uses and limits of resource exploitation if escalating conflicts and resource degradation are to be avoided.

The attitudes of community and industry to the use of biological resources should change from the 'maximum yield' approach to one of 'ecologically sustainable', which recognizes the need for conservation of biological diversity and maintenance of ecological integrity. Integration of management regimes within and between different sectors to meet environmental, economic and social objectives must be realized in order to achieve sustainable development.

Integrated Approach

Integrated policies will also provide the opportunity for all the people to accept responsibility for their actions and the impact they may have on biological diversity. The development of integrated policies for managing the coastal resources is necessary:

- to coordinate activities within and between all levels of government;
- to ensure that full social and environmental consequences (and costs) of development activities are considered;
- to ensure that the public interest is properly taken into account.

ASSIGNING ECONOMIC VALUE TO BIODIVERSITY

Biodiversity refers to the diversity of life in all its forms and all its level of organisation, not just plant, animal and microorganism species. At its most elemental level, biodiversity, encompasses the varied assemblages of organic molecules that comprise the genetic basis of life. On the other end of the spectrum there are biomes-the vast stretches of tundra, desert, forest, ocean, etc. In between come-population, race, sub-species, community, ecosystem.

Why is Biodiversity Important?

For many people the very questioning of the worth of biodiversity is illicit. They would argue that humankind has a moral obligation to conserve biodiversity, an obligation that comes with the fact that humans have the capability to destroy much of that biodiversity. Others find a religious support for such a view there is some stewardship responsibility on behalf of some deity.

One problem with these moral views is that they often conflict with other moral views about, say, the right to earn a living, the right to have access to basic needs such as food, cloth and shelter, and so on. If conserving biodiversity conflicts with those rights, then some 'meta-ethical' principle is required for deciding which moral view should prevail. One aspect of the process of changing popular perceptions about biodiversity resources is to show that the sustainable use of biodiversity has positive economic value and that this economic value will often be higher than the alternative resource uses which threatens biodiversity.

Valuing Biological Diversity Conservation

The issue at hand concerns the measurement of natural capital, great difficulties arise in assigning values to natural capital precisely because the market in which it is traded are limited. In the past few decades a number of techniques have been developed to place monetary value to the benefits of conserving an area for biodiversity. Some rely on market prices of related goods and services both to value benefits and to estimate costs, while other rely on survey based approaches to infer values.

Hedonic Pricing Method

Property values are affected by a number of variables including environmental quality. After excluding other variables, including environmental quality, the residual price difference can then be ascribed, at least theoretically, to differences in environmental quality, *e.g.*, the increased value of property located next to natural areas. According to a report by Nelson (1982) traffic noise, measured in Leq (equivalent continuous sound level), a one unit change produces property price depreciation of 0.5-1.0%.

Ravel-cost Method

This approach looks at the pattern of recreation use of a natural parks and uses this information to derive a demand curve to estimate the total amount of consumer's surplus. For example, travel behaviour reveals that Costa Rican visitors are willing to pay USD 35 per household to visit a tropical rainforest site in Costa Rica. Foreign visitation is likely to be worth far more than domestic, as foreign visitors have higher travel costs.

Contingent Valuation Method

The Contingent Valuation Method (CVM) bypasses the need to refer to market prices by asking individuals explicitly to place values upon environmental assets. This is also referred to as an expressed preference method. An interesting advantage of the CVM approach is that it can, in theory, be used to evaluate resources, that people have never visited personally *e.g.* the Antarctica which people are " Willing to Pay" to preserve but would not in general ever want to visit.

Consumptive Benefit Method

There are a number of products which can be harvested on sustainable basis and marketed. For example, a probable number of higher plant species, which are widely used as basis for pharmaceutical drugs, is some 500,000. According to Pearce and Moran (1994) economic value of these plants can be approached by looking at:

(i) the actual market value of the plants when traded,
(ii) the market value of the drugs of which they are the source materials,
(iii) the value of the drugs in terms of their life-saving properties, and using a value of a 'statistical life',
(iv) the lost pharmaceutical value from disappearing species.

As there is absence of 'global markets' in the benefits of biodiversity, the developing countries face major problems of appropriating the global benefits of sustainable use of biodiversity. As long as these global values cannot be captured by host countries, biodiversity will be a risky investment in many contexts.

COMPLEXITY OF ENVIRONMENTAL VALUATION

Complexity of environmental valuation also arises due to the multi-ple functions of being a source of raw materials and energy, being a sink for assimilating man-made wastes and providing other services such as recre-ational/tourism services, storage of genetic diversity and scientific and educational benefits.

The following classification is useful:

- Direct use values are derived from the economic uses made of the natural system's resources and services. Examples of these are outputs such as timber, game and recreation from forests or fish and scuba tourism from coral reefs.
- Indirect use values are the indirect support and protection provided to economic activities and property by the resource system's natural func-tions or environmental services. Examples of these are watershed protection and soil erosion prevention provided by forests, and beach sand and mooring facilities protection provided by coral reefs.
- Non-use values lie in the special attributes of the natural system as a whole, its cultural and heri-tage uniqueness; it includes both existence and option values. Existence values reflect public goods which can be enjoyed by more than one consumer without decreasing the amounts en-joyed by others such as clean air, beaches and forests. The existence value is the utility that consumers derive from just knowing the public good exists. A way of measuring that utility would be to measure the willingness to pay or the contingent value assigned

if the public good were to disappear. Option values reflect what current generations wish to bequeath for future generations to inherit. They imply both an ethi-cal commitment to sustainabilty for the children of our children and in a shorter time-frame the maintenance of options to solve current prob-lems. Examples of option values have been devel-oped for forests relating to biodiversity and the search for cures of cancer and AIDS-"if forest were to disappear then the options to find such cures will vanish" or "as long as the forest is pro-tected there is the option to find the cure".

Direct Effects Valued on Conventional Markets Some methods are directly based on market prices or productivity. This is possible where a change in envi-ronmental quality affects actual production or pro-ductive capability.

Change-in-Productivity: Development projects can affect production and productivity positively or neg-atively. The incremental output can be valued by us-ing standard economic prices where available. Loss-of-Earnings. Environmental impacts can signifi-cantly affect human health. In theory, the value of health impacts should be determined by the willing-ness to pay of individuals to maintain their health. In practice, one uses earnings lost upon early death, dis-ease or job absence. This approach is used in highway and industrial safety, and in air pollution studies.

The "implicit value of human life" approach is reject-ed by many as dehumanizing since human life can be said to have infinite value. However, society, govern-ment regulations, insurance companies and judicial courts implicitly and explicitly place finite values on human life and health. This is a necessity reflecting limited resources to be allocated for health expendi-tures. The relatively high level of health expenditures in Cuba would indicate a high implicit value of hu-man life and health. However, one can also discern a political motivation and its attendant benefits behind it.

Preventive Expenditures: Individuals and governments invest in prevention measures to avoid or reduce un-wanted environmental effects. Environmcntal dam-agcs, arc oftcn difficult to assess, but historical infor-mation on preventive measures and their costs may be interpreted as a minimum value for the expected benefits that the preventive measures seek. If it is found, for example, that there is a historical pattern of under investment in Cuba for natural disaster planning and prevention; then, it can be inferred that the benefits expected from such preventive measures have been very low. This conclusion would in turn imply that the loss of human life would be assigned a relatively low value. While this possible conclusion may appear to conflict with the high political priority for health expenditures, it can be observed that the level of damages or loss of lives caused by natural di-sasters cannot be easily attributed to government in-action and thus the political cost may be easily ex-plained away.

Potential Expenditure

Valued on Conventional Markets

Replacement Cost: Simply, the costs that would have to be incurred in order to replace a damaged asset. The estimate is not a measure of benefit of avoiding damage since the damage costs may be higher or low-er than the replacement cost.

Valuation Using Implicit (or Surrogate) Markets

Sometimes one must use market information indi-rectly. Approaches to be considered are the travel cost method, the property value approach, the wage dif-ferential approach, and uses of marketed goods as surrogates for non-marketed goods. Each technique has its particular advantages and disadvantages, as well as requirements for data and resources. One must determine which techniques might be applica-ble to a particular situation.

Travel Cost: This approach measures the travel cost which reflects the willingness of consumers or users can serve to measure the benefits produced by recreation sites (parks, lakes, forests, wilderness). It can also be used to value "travel time" in projects dealing with fuelwood and water collec-tion.

Property Value: This valuation method is based on the general land value approach and can determine the implicit prices of certain land areas. The property value approach can help analyze willingness to pay for properties with different pollution levels and infer the implicit cost of pollution. The method compares prices of houses in affected areas with equal size and similar neighborhood characteristics elsewhere in the same metropolitan area.

Valuation Using Constructed Markets

Contingent Valuation: When society's preferences as revealed in market prices are not available, the con-tingent valuation method tries to obtain information on individual preferences by posing direct questions about willingness to pay. It basically asks people what they are willing to pay for a benefit, and/or what they are willing to accept by way of alternate compensation to tolerate an environmental cost.

This process may be achieved through a direct questionnaire/sur-vey. Willingness to pay is difficult to measure and de-pends on the income level of the sample subjects, and involves problems of designing, implementing and interpreting questionnaires. While its applicability may be limited, there is now considerable experience in evaluating the quality of supply of potable water and electricity services.

Artificial Market: Such markets can be constructed for experimental purposes, to determine consumer willingness to pay for a good or service. For example, a home water purification kit might be marketed at various price levels, or access to a game reserve may be offered on the basis of different admission fees, thereby facilitating the estimation of values placed by individuals on water purity or on recreation facilities.

THE DISCOUNT RATE

Discounting is the process by which costs and bene-fits occurring in different time periods may be com-pared. The discount rate to be used has been a gener-al problem in cost-benefit analysis, but it is particularly important with regard to environmental issues, since some of the associated costs and benefits are very long-term or irreversible in nature. In stan-dard analysis, past costs and benefits are treated as "sunk" and are ignored in decisions about the present and future.

Future costs and benefits are discounted to their equivalent present value and then compared. In theory, in a perfect market, the interest rate re-flects both the subjective rate of time preference (of private individuals) and the rate of productivity of capital.

Higher discount rates may discriminate against fu-ture generations. This is because projects with social costs occurring in the long term and net social bene-fits occurring in the near term, will be favored by higher discount rates. It is often argued that discount rates should be low-ered to reflect long-term environmental concerns and issues of intergenerational equity. However, this would have the drawback that not only would eco-logically sound activities pass the cost-benefit test more frequently, but also a larger number of projects would generally pass the test and the resulting in-crease in investment would lead to additional envi-ronmental stress.

Many environmentalists believe that a zero discount rate should be employed to protect future genera-tions. However, employing a zero discount rate is in-equitable, since it would imply a policy of total cur-rent sacrifice, which runs counter to the proposed aim of eliminating discrimination between time periods-especially when the present contains wide-spread poverty.

In the case of projects leading to irreversible damage (*e.g.*, destruction of natural habitats, etc.), the bene-fits of preservation may be incorporated into stan-dard cost-benefit methodology using the Krutilla-Fisher approach. Benefits of preservation will grow over time as the supply of scarce environmental re-sources decreases, demand (fueled by population growth) increases, and possibly, existence value in-creases.

The Krutilla-Fisher approach incorporates these increasing benefits of preservation by including preservation benefits foregone within project costs. The benefits are shown to increase through time by the use of a rate of annual growth. While this ap-proach has the same effect on the overall CBA as low-ering discount rates, it avoids the problem of distort-ed resource allocation caused by arbitrarily manipulating discount rates.

CONCLUSIONS

In order to achieve economically sustainable manage-ment of natural resources and environmental protec-tion, one must effectively incorporate environmental concerns into decision making through the EIA pro-cess.

This presentation has reviewed concepts and tech-niques for economic valuation of environmental im-pacts within EIA procedures. The process of internal-izing these environmental externalities can be facilitated by making rough qualitative assessments early on in the project evaluation cycle-the advantages of which would include:

- early exclusion of alternatives that are not sound from an environmental point of view;
- more effective in-depth consideration of those al-ternatives that are preferable from the environ-mental viewpoint; and
- better opportunities for redesigning projects and policies to achieve sustainable development goals.

In order to fully reflect society's values and preferenc-es about environmental values, non-market methods of estimation can be used and will enhance commu-nity participation through well designed and admin-istered questionnaires and surveys. Research and training about EIA and embedded economic analysis methods is needed in Cuba. As developing countries learn to successfully apply EIA methodologies the goals of sustainable development will become more attainable.

INTERACTIONS AND INTERMEDIATE CONDITIONS

The forests described above are stereotypes that represent the ends of the range of variation. In practice, many forest types are intermediate in some respect, although their dynamics and composition can be related to the stereotypes. Some examples illustrate the range of possibilities.

- Forests in which partial stand destruction is commonplace can develop into complex mosaics of tolerant and intolerant species. An example is the *Pseudotsuga menziesii* forests of the Pacific North-west, where saplings and underwood are frequently burned but the overstorey is unaffected. Moreover, light fires tend to be patchy, burning some places but leaving other patches untouched. This allows the overstorey of *Pseudotsuga menziesii* to be infiltrated by *Abies amabilis,* which develops as patches in a mosaic with groups of younger *Pseudotsuga menziesii* in larger burned patches.
- Complex interactions can develop between disturbance regimes. For example, in the southeastern USA *Pinus* forests were maintained naturally by frequent fires. However, where stands happen to remain unburned, *Quercus* and other broadleaved species colonize, form an underwood and eventually dominate the canopy. Since the broadleaved

stands are relatively fireproof they tend to remain undisturbed, thereby generating a patchwork of broadleaved and *Pinus*-dominated stands whose pattern is determined by disturbance history. This capacity for two forest types to develop on one site type is inherent in any region where disturbance becomes less likely as succession proceeds.

- The characteristics of *Fagus sylvatica* in central Europe determines much of the forest pattern. Its capacity to coppice enables it to form distinctive low scrub woodlands with *Sorbus aucuparia* near the treeline, where wind and snow maintain a chronically disturbed environment. In this case, the *Fagus sylvatica* is effectively undisturbed. Elsewhere, in steep slopes in gorges, the stands are chronically destabilized by the fall of large trees growing on insecure rootholds, and this breaks the dominance of beech and enables a mixture of tree species to survive, including *Tilia, Ulmus* and *Fraxinus.*

Disturbances have long-term effects on forest composition. For example, drought rarely destroys stands entirely but confers an advantage on species that survive droughts. Ultimately this is expressed as a distinctive assemblage on drought-prone sites, such as outcrops and convex slopes with thin soils. In this instance, disturbance tends to generate adapted forests, which thus reduces the incidence of disturbance. Alternatively, species that depend on disturbance may both invite it and be ready to survive. The best examples are the fire-dependent *Pinus banksiana* and *Pinus contorta* forests, which are inherently inflammable, particularly those that exhibit serotiny, where long-lived cones release their seed only after fire and which falls into a seed bed free of litter and competing vegetation. A more extreme example is seen in the *Pinus mugo* of avalanche tracks, where repeated disturbance has provoked the evolution of a species that can withstand the disturbance.

The multitude of interactions between trees and disturbance regimes is expressed as different assemblages on different site types and as a range of successional states on each site type. Disturbance should be seen as an integral component of the forest type, not an external destructive force. Collectively, disturbances enhance regional diversity by creating distinctive forest types, maintaining a range of successional states, and by enhancing the amount of edge in the landscape.

Influence of People

The relationships described so far have dealt with natural conditions. In practice, most forest types are overwhelmingly influenced by people:

- Secondary successions to forest on land previously cleared are a response to severe disturbance, *i.e.* the complete loss of tree cover and the alternative use of land for cultivation or pasture. Pioneer trees are almost always intolerants. The pattern of secondary succession depends on seed sources, so that colonization may be extremely patchy. Colonization often takes place in waves, for example when initial colonization by wind-dispersed trees is followed by bird-dispersed trees once perches have been created.

- Coppice systems enable shrubs and intolerant trees to form a higher proportion of a stand than they would naturally maintain. Despite the constant disturbance of felling, individual trees can sprout again indefinitely and thus live well beyond their natural span. In terms of individuals and small-scale patterns, coppices are probably more stable than the natural forests from which they were derived.
- In woodland pastures, regeneration is inhibited by grazing and browsing, although established individuals have little competition and can expand to great sizes. Moreover, the constant lopping of branches for fuel and fodder, followed by regrowth from the pollard, enables individual trees to grow to great ages. Here, too, the effect of management disturbance is to generate unnaturally high stability.

Finally, there is the problem of *Quercus petraea* and *Quercus robur* in Europe. These are long-lived intolerant species that have been abundant in the pollen rain over very long periods, which implies that the forests were considerably disturbed, albeit at long intervals.

They have been maintained at unnaturally high levels by traditional management, which favoured them as timber trees in coppices and wood-pastures.

Under natural conditions it is possible that oaks were maintained by the high level of browsing and grazing imposed by populations of large herbivores, such as deer, horses and cattle. In fact, it has been proposed that natural forest was more like savannah before prehistoric hunters reduced and domesticated the larger herbivores.

However, this seems far from proven, although the possibility remains that herbivore populations maintained a controlling presence on the structure and composition of some types of temperate forest, particularly perhaps those on wet ground and heavy fertile soils (where ground vegetation would be vigorous) and steep dry slopes (where tree growth might be poor).

THE BIOLOGICAL MICRO-ENVIRONMENT

Organic matter, microscopic and macroscopic organisms (*e.g.*, fungal hyphae and invertebrates), detritus from fungi and animals, and bacteria, and biological exudates, all assist in stabilizing soil structure. The role of each part of the biomass differs according to its size. Broadly, large aggregates greater than 250 m m diameter (macro-aggregates), are stabilized by their inherent physical structure, wetting and drying cycles, and organic matter. Micro-aggregates (< 250 m m) are stabilized by live or dead roots, fungi, invertebrates and microorganisms.

The populations of soil organisms of all sizes are linked functionally through their roles in the degradation of various forms of organic material. The latter includes live and dead plant material and other live or dead organisms. This shows that animals such as nematodes and some fungi feed directly on live plants while other fungi and bacteria feed predominantly on litter. Earthworms and other large invertebrates create, and inhabit, burrows and pores, and are very mobile. The most notable of these are termites, which are divided into three groups according to the structure of their nests: those that build mounds (a) above ground, (b) on the soil surface, and (c) below ground. Small arthropods, microfauna and fungi live mostly in larger voids and in association with roots. Foster (1988) reviewed the location of the various types of soil-dwelling organisms and found that fungi, which constitute about 80% of the biomass in many soils, tend to be restricted to the rhizosphere of roots, to larger pores between aggregates and to the surface of aggregates. Bacteria, by contrast, are found on roots in the rhizosphere, in small colonies in the larger micropores, within aggregates and on and within cell debris.

Organic Matter

Both plants and animals provide inputs of organic matter to soils. Once within the soil organic residues can be distinguished on the basis of their chemical structure (*e.g.*, old lignified humic substances that degrade slowly), by their source (plant or animal) or by location.

The standing crop of litter in semi-arid grasslands is usually more than 3 t/ha and in temperate dry steppe may exceed 11 t/ha. There has been much debate about the relative contents of organic matter in tropical and temperate soils. Within those wet-and-dry climates that have hot summers assisting rapid decomposition, there is no evidence of inherently lower levels of organic matter in the tropics than in comparable temperate regions. Kowal and Kassam (1978) and Juo and Payne (1993) review the role of organic matter in tropical soils. Here, it is sufficient simply to state that organic matter has various interrelated effects on soil fertility. In particular it should be noted that both chemical and physical effects are of relatively great importance in the soils of the semi-arid tropics because these generally have low cation exchange capacity. The relative importance of litter (crop residue) and manure as inputs of organic matter, varies between cropping systems and spatially within a system. Here, most of the above-ground crop residue is fed to animals, but an equal amount of below-ground crop material enters the soil organic matter pool.

Where alley cropping and agroforestry are practised, values are more variable, but possible inputs could be very significant where the trees, from which the litter is taken, are grown away from the annual crops. If, say, two-thirds of the leaves from leguminous trees are harvested annually, litter values will be substantially higher and the material of better quality than the leaf and stem residue from an annual crop likely to be recycled in the field. Some qualifications, however, should be made. Tree root material is not available for decomposition in the crop field unless it is spatially overlapping (*e.g.* as an intercrop), in which case the trees will compete with the crop for soil nutrients, water, light and space.

The proportion of animal and human manure used on cropland is more variable. Some farmers have developed stable systems which strongly emphasize the use of animal manure on crops. For example, Norman *et al.* (1982) describe how farmers in northern Nigeria managed to apply 4 t/ha of manure to their heavily-cropped crop land though they had only 3 cows each. Many other farmers do not ensure adequate recycling, either because they are more concerned with livestock management or they do not know the importance of maintaining a 'zero nutrient budget' to replace nutrients removed by the crop. For example, Norman *et al.* (1982) also describe farmers with 10 cows each, who applied only 1.9 t manure/ha to their crops.

Within a cropping system, manuring practice varies with location. There is transference towards the centre of the system. On traditional farms, the area near the household or village is highly fertilized with human and animal manure while more distant fields receive little or no organic matter. Fussell (1992) describes such a traditional 'ring' farming system in semi-arid west Africa. Here, if houses are thatched, the village needs rebuilding or moving every 2 to 4 years. Moving takes advantage of the fertility gradient. Where the huts are not moved, the fertility gradient becomes steeper with time. Rather than trying to even out fertility by labour-intensive transport of manure, farmers vary the cropping of the fields. Continuous cropping of millet is sustainable close to the hut or village where there is plenty of human and animal manure but crop rotations are essential at the periphery.

7

Forest Trees from Biotechnology

Most of the biotechnologies used in applied forestry today fall into the categories of tissue culture and molecular marker applications. Advances in these two areas of biotechnology have all been logical progressions and refinements of techniques that have been used in plant and tree improvement for decades, if not centuries, to produce clonal lines or varieties. Most of the popular concern about GM plants, or about the use of any of the other techniques discussed in this article, derives from concern about products grown as food for human consumption. This will rarely be a direct issue in forestry, although careful evaluation will be necessary in the case of multi-use tree species that provide non-wood forest products (NWFPs).

The argument that the genetic changes made in GMOs are "unnatural" has been raised as a concern. It is, however, difficult to say that any particular technique in itself can pose increased biological risk. It is the gene products that may pose a lower or higher risk, irrespective of the technology used to obtain them. To evaluate possible risks fully, in-depth knowledge of the particular genetic transformation is necessary, *e.g.* the product produced, where the transformation occurred, what "promoter" genes were used and the reliability of the expression over time (Gutierrez, MacIntosh and Green, 1999). This may be particularly true when more than one or two transgenes are incorporated. In such cases there may be requirements for longer field testing, environmental assessments and more caution in deployment (Burdon, 1999). In addition, mutation caused by gene transfer and tissue culture (somaclonal variation) can alter the function of non-target genes in ways that might take some time to detect. These, however, are not genetic issues, but a matter of having appropriate genetic testing and evaluation programmes in place prior to the release and use of GM (or even conventionally bred) trees in managed landscapes.

ISSUES ASSOCIATED WITH SPECIFIC TYPES OF TRAITS

Herbicide resistance. Herbicide resistance in poplars is probably the best-developed GM technology in forest trees. The first concern with herbicide-resistant GM plants is evidence of the development of resistance in weed populations. The risk may be substantially less in forestry than in agriculture, as herbicides are only used for a short time and with fewer applications during the early period of plantation establishment. Furthermore, total weed control is not necessary in forest tree plantations, so there is less selection pressure for resistance to weeds. The other concern is cross-breeding with wild populations. While this is possible, it is likely that wild populations that do hybridize with GM trees would have little selective advantage unless adjacent wild populations are treated with herbicide, which is rare in forestry. The main question is then the effect on the genetic structure of local wild populations if cross-breeding does occur (*e.g.* in *in situ* conservation areas) and the acceptability of this risk to local resource managers. If this is a concern, reduced flowering or sterility transgenes may be required, along with herbicide resistance in GM trees.

The introduction of herbicide resistance through GM technology is likely to be the most feasible and the most frequently applied genetic modification in trees. Nevertheless, it is only likely for a few well-developed species in certain situations, such as in intensive poplar fibre farms where herbicides are sometimes an approved management tool.

Reduced flowering or sterility. Reduced flowering in forest trees may be desirable in order to steer the products of photosynthesis into wood production, rather than reproductive tissues. However, since such reallocation of resources has not yet been well quantified, the main justification for reduced flowering or sterility development is when it substantially reduces the gene flow to wild adjacent populations of the same species. This may help promote greater acceptance of intensive GM tree plantation forestry adjacent to natural forests. Although research on flowering mechanisms is under way (*e.g.* Skinner *et al.*, 1999), the stability of sterility-gene expression over time will have to be confirmed in field trials that reflect expected rotation lengths.

Insect resistance. The development of GM crops that are insect resistant is now common, but these crops also create some of the most complex ecological situations that need to be addressed. The first issue of concern is the possible toxicity of the compounds produced in GM insect-resistant plants when they are grown specifically for human consumption or for animal feed as part of associated food chains. Second, there are ecological concerns of cross-breeding with wild relatives and the evolution of resistance in the pest populations, as with herbicide resistance. An additional and serious problem in forestry is that the long generation time of trees allows for many generations of insect populations to challenge a resistance mechanism.

The most developed GM approach for insect resistance in forestry, as in agriculture, has been the use of genes from a natural insect pathogen, *Bacillus thuringiensis* (*Bt*). Poplars are again among the tree species in which the technology is most advanced (TGERC, 1999). Research and development of other compounds is under way to reduce reliance on the relatively narrow group of natural *Bt* toxins (ffrench-Constant and Bowen, 1999). Because of the complex ecological ramifications and public concerns surrounding GM insect-resistant plants, high levels of scientifically sound laboratory and field testing will be required.

Wood property chemistry. It is now technically possible to alter genetically the chemistry of lignin in trees for easier and more environmentally friendly pulping. Genes that are important to the pathway of lignin development in wood have been modified to produce unique wood composition in very young trees (Lapierre *et al.,* 1999). However, genes of relatively large effect on wood chemistry have also been found naturally (*e.g.* a major recessive gene in loblolly pine [Ralph *et al.,* 1997]), so wood properties can also be modified through conventional selection and breeding. Therefore, many of the same ecological and economic considerations need to be taken into account with or without genetic transformation technology. Two important questions that remain in developing lignin-modified varieties or clones are how economically valuable plantations using such trees would be and whether altered wood may show susceptibilities to environmental stresses. If the answer to the latter is affirmative, it is again likely that reduced flowering or sterility genes would have to be incorporated into material used on a large scale when it is important that there be minimal gene flow to adjacent wild populations (*e.g.* in *in situ* conservation areas).

Can genetically modified trees be safely deployed in the environment?

The main ecological concerns with GM plants is the issue of gene exchange with wild populations - an issue already being considered in forestry with regard to seed orchard seed from conventional breeding programmes. In most of the situations for which GM trees will be considered, the plantation species will be an exotic one, so gene exchange would not be a factor; but in cases where it could occur, reduced flowering or sterility is likely to be a basic requirement. Much has been and will continue to be learned from agricultural research on transgene flow from GM crops to their wild relatives. Investigations into these specific questions are under way in forestry (DiFazio *et al.,* 1999). Although exotic plantation species may have limited risk of cross-breeding *per se*, the release into the environment of a GMO in most cases poses less of an ecological risk than the introduction of an exotic species. The introduction of an exotic species is a release into the environment of an entirely different genetic constitution, rather than the introduction of a single gene into a native species. The current regulations of the Forest Stewardship Council (FSC, 1999) prohibit the use of GM trees, but suggest that "the use of exotic species shall be carefully controlled and actively monitored to avoid adverse ecological impacts". This recommendation largely runs contrary to the argument made here that the biological risks and economic realities of all forest management options need to be evaluated objectively. On the other hand, a large genetic change made to the overall fitness of a native species and released without adequate consideration of local environmental risks is not appropriate either.

Clearly, potential risks must be balanced with benefits in all types of improved forest trees, considering the deployment schemes that will be used in different locations and at different times. For example, clonal varieties of fruit trees have been used in large continuous orchards for hundreds of years, and this deployment and management approach is considered by most people to be ecologically and economically acceptable, as it presents no unmanageable biotic or abiotic challenges. In plantation forestry for wood production, clonal blocks may be appropriate in some situations, such as poplar plantations, while pure or family-mix planting patterns are likely to be appropriate for most other forestry situations. In any event, deployment strategies must be designed to minimize the risks of economic losses of the stand as well as future biological losses, such as the development of resistance in pest and disease populations (Roberds and Bishir, 1997). At a minimum, therefore, a few dozen well-tested GM clones will be necessary to meet the typical production and diversity requirements necessary even in intensive forestry programmes.

To help address many of these concerns, a number of countries have developed regulations and restrictions specifying the requirements of confined field testing that are needed before commercial release (OECD, 2000). These requirements, which are necessary to reduce biological and economic risks, will undoubtedly continue to evolve. In addition to national laws and regulations, there are also broader international agreements on biosafety, *e.g.* the Cartagena Protocol on Biosafety (Convention on Biological Diversity, 2000).

EQUITY OF ACCESS TO GENETIC MODIFICATION BIOTECHNOLOGIES

Private investors have taken the lead for most investment in modern biotechnology, and in so doing have also had to manage the associated economic risks. In many situations such investment risks are protected by patents. An obvious concern is, therefore, that access to GM crops and trees will be controlled almost entirely by private corporations. Agreements for the use of techniques or material could be prohibitively expensive, preventing the use of the technology where it would be of value. In the past, governments were the main investors in breeding and molecular biology research. Their role may have to expand once again in this regard, in order to provide a flow of material and information that can be developed and shared by both private and public institutions (Santos and Lewontin, 1997), as opposed to being controlled largely by private companies. Nevertheless, financial constraints, rather than available technical knowledge, are probably the largest challenge ahead for applying modern biotechnology in less developed countries, although it has been shown, *e.g.* by the Agricultural Biotechnology Support Project (www.iia.msu.edu/absp/), that arrangements through donations are possible.

Even if financial considerations can be addressed in the research phases of biotechnology, moving the technology from the development stage to operational reality provides a new set of technical challenges with additional expenses (Polonenko, 1999). It is likely that only a few economically important species will warrant such additional investments, and it is on these economic situations and realities that the debate should focus.

Public acceptance of genetic modification

As has been alluded to previously, in terms of biotechnology research, one of the most important economic considerations separating forestry from agriculture is that trees require substantially more time for proper evaluation under both laboratory and field conditions. Rapid generation turnover in crop species has allowed genetic modification technology to develop quickly. Although this has been desirable from an economic point of view, the ecological ramifications have been far more difficult to study, as complex experiments must be carried out over several years. Public and government acceptance of GM plants is now probably as dependent on biological risk assessment and risk issue management (Leiss, 1999) as it is on technical developments or economics alone.

For GM trees, there is clearly a need to provide scientifically sound, neutral and intelligible information to the public, so that when and if GM trees become commercially available, informed decisions can be made about their release into the environment. Foresters are generally very aware of the need for public involvement in many other areas of forest management (*e.g.* chemical applications in forests, game management, felling patterns and plans, allowable annual harvesting levels), so decisions concerning GM trees may simply be an additional issue to be managed in the process.

CONCLUSIONS AND FUTURE CHALLENGES

The main issues that should remain relevant to the use of any modern biotechnology in forestry, now and in the future. The following are some summary points and conclusions:

- The development time, from the laboratory to the field, for all new types of tissue culture or GM trees will be substantial, usually in the order of a decade or more. While this may provide forest scientists and managers with time to evaluate many of the issues being faced in agriculture, the economic realities of relatively long generations will continue to be a major challenge to investors in biotechnology in forest trees. It appears that genetic modification will therefore become a reality only for particularly novel and valuable traits in short-rotation species in intensively managed plantations. Herbicide resistance is likely to be the first trait for which genetic modification will become economically viable, but even this may be several years away from large-scale use.

- All modern biotechnologies require large research and development investments. The allocation of funds, through either private or public agencies, needs to achieve a balance between building scientific capabilities and knowledge and supporting more applied, well proven forestry technologies (Burdon, 1994). In this regard, the investment and use of any biotechnology needs to be assessed on a case-by-case basis, considering the specific technology in the specific ecological, political and economic environment.
- The public must be confident that regulatory frameworks are in place so that commercial development of biotechnology does not take precedence over biosafety issues. However, the development of useful biotechnology in forestry should not be put at risk with extraordinary regulatory compliance costs that are driven by unwarranted concerns.
- Government participation in the development of biotechnology appears necessary to keep some flow of material and knowledge in the public domain. Without this, the costs of using privately owned genes could seriously limit further interest and development of genetic modification technology in trees. Concerns about monopolies arise with any new products, but governments can step in to deal with substantial inequities or corporate actions that are not in the interest of society as a whole.
- Even if GM trees are not used on a wide scale in the future, research and development in GM technology will provide a great deal of information about gene action and regulation, which could be of long-term value to conventional breeding programmes.
- Other forest management decisions with potentially more serious ecological consequences, including large-scale species introductions or inappropriate use of provenances or improved trees from even conventional breeding, need to be evaluated by foresters, managers and regulatory agencies in the same way as the products of modern biotechnology are.
- Geneticists have always attempted to create genetic options for the future. If we look carefully at where these current and future biotechnologies are heading in forestry, it is obvious that the research and development will largely be used in support of advancing clonal forestry. Therefore, it is in the context of appropriate clonal forestry that most of the issues need to be examined and evaluated.

From a genetic perspective, concerns that biotechnology is "unnatural" ignore the dynamic changes in the genetic code that occur within and across species genomes through modification of transposable genes or elements by virus vectors and through mutation.

This article has attempted to point out that humans have already had a huge effect on the genetic structure of many plants and animals, using many different technologies which will continue to develop and change over time. However, while those in the forefront of any technology will promote its potential benefits, in the end it will be the economic and regulatory systems of governing bodies at the national and global levels that must evaluate the technology's relevance and appropriateness.

CHANGES OUTSIDE THE FOREST SECTOR

The example in the last paragraph, which would be applicable to many other developing countries, indicates the unpredictability of changes and their impact on forestry. Even with the best forecasting tools that were available three decades ago, very few of the changes described could have been visualized. Sector analysis focused on measurable values and on estimating demand and supply, largely based on projecting historical trends. The key parameters considered in assessing the direction of forestry development were limited to population, urbanization and changes in income, supply and prices of substitutes. Fundamental system-wide changes in the economy as a whole could not be captured by such forecasting techniques; thus many changes visualized then were rather off the mark. Many intentional efforts to bring about changes in forestry had little impact, while most changes were unintentional and not necessarily the outcome of planned efforts.

Since changes in the rest of the system can alter development scenarios drastically, it is important to understand how system-wide changes take place and what may be done by foresters to take advantage of the emerging opportunities. For convenience, the changes could be categorized as economic, institutional, environmental or technological.

ECONOMIC CHANGES

Among the major driving forces of change is the increasing economic integration and interdependency of countries. The rejection of centralized planning and the adoption of liberal economic policies have opened up economies, with the market mechanism becoming the most dominant determinant of change. Some of the impacts that are relevant to forestry include:

- Opening up of local markets for imported forest products;
- Increased transnational investment in forestry, especially logging, plantation development and wood processing, largely based on perceived comparative advantages;

- Improved opportunities for marketing unique products (*e.g.* products from medicinal and aromatic plants, which were earlier used only locally; ethnic foods) and services (*e.g.* ecotourism);
- probable adverse impacts on local industries that fail to adapt to the changes and thus lose the comparative advantages that they had in a protected environment.

The impacts of globalization, which results in the easy movement of capital, technology, goods and services across national boundaries, are sometimes difficult to assess. The set of criteria previously used for measuring comparative advantage for investment in forestry (*e.g.* nearness to markets and raw material supplies; quality and quantity of raw material) is expanding to include very different criteria (*e.g.* need to reduce pollution; degree of openness of economies; barriers to trade).

Institutional changes

While globalization is increasing the interdependency of countries and societies, more pluralistic institutional arrangements are emerging (FAO, 1999). Improved access to information is changing the centre of power and authority and, as part of a larger process of devolution of administrative responsibilities with wider emphasis on participatory approaches, local communities are playing a lead role in resources management decisions. The involvement of the private sector in forestry, including forestry research, is increasing (Enters, Nair and Kaosa-ard, 1998).

Furthermore, increasing awareness of forestry issues has brought civil society to the forefront of influencing forestry decision-making (FAO, 1998a). Improved access to information, the media's new consciousness of the public and the emergence of democratically functioning and transparent institutions, as well as impartial and just mechanisms to redress grievances, have all enhanced the ability of civil society to intervene in critical issues of public interest.

Increased concern for environmental protection

Undoubtedly, the most important development affecting forestry in recent years is the increasing awareness of the environmental issues relating to forest resources management, with initiatives at all levels (including the global-level Convention on Biological Diversity, Convention to Combat Desertification and Framework Convention for Climate Change, which were initiated by the United Nations Conference on Environment and Development [UNCED], and country-level efforts to revise forest policies) giving emphasis to environmental benefits. Outcomes of this awareness include ongoing efforts to develop and refine criteria and indicators for sustainable forest management, certification and labelling, codes of logging practices and the extension of protected areas.

Technological changes

Developments in science and technology, both within and outside the forestry sector - more particularly the latter - have had a significant impact on forestry (FAO, 1998b). Advances have included enhanced utilization potential for several species, reduced raw material requirements through improved efficiency in processing, substitution and lowered emission of pollutants. Recycling has increased and, as in the case of paper, could drastically reduce the demand for wood (Abramovitz and Mattoon, 1999). Research on biological aspects of forestry, including ecosystem processes, have enhanced the tools and techniques available for sustainable management of plantations and natural forests. Genetic selection, tree breeding and rapid multiplication techniques, coupled with refined site management practices, have enhanced the potential to realize higher productivity from tree plantations.

Developments in information and communication technologies are unleashing unprecedented changes. Resource monitoring capabilities, including imaging resolution and interpretation, have improved considerably. Forest cover changes, pest and disease infestations and outbreaks of fire are more easily monitored, and further improvements that will enable wider easy real-time access to resource information are a distinct possibility in the near future. The spread of e-mail and Internet access is improving the ability to communicate and thus to share information and experience. Distance is no longer a critical barrier for interaction among people. This is altering the structure and functions of organizations, with purpose-oriented networks becoming a dominant institutional arrangement (Nair and Dykstra, 1998), while the traditional vertically structured organizations are slowly fading out (EIU, 1997).

EXPLAINING THE CHANGES: LOOKING BELOW THE TIP OF THE ICEBERG

Given the potential of the economic, institutional and technological changes, why has the forestry situation failed to improve, or even worsened, in several countries during the past few years? Why is change rapid in some countries but painstakingly slow in others? In some cases efforts to bring about economic, institutional and technological changes are not possible to sustain, as the system as a whole in some way rejects them. The adoption of change seems to be largely dependent on system-wide developments and, more particularly, structural shifts in economies.

Structural changes in economies

Historically, agriculture (including animal husbandry) has been the main source of livelihood for most people in the early stages of economic development. Economic progress is largely dependent on the production of surpluses over and above consumption, and their transformation into other goods and services through either trade or investment. As populations grow, production is increased by bringing more land under cultivation or by intensifying agriculture through yield-enhancing technologies. Under the extensification scenario, forest cover tends to decline, as most agricultural expansion takes place on forest land. In many countries, agriculture remains the mainstay of the economy, and consequently population increase implies additional forest clearance for cultivation. Shifts away from land-based activities occur largely with the emergence of other sectors.

In several countries, the discovery of oil and natural gas has led to structural shifts in the economy (Wunder, 2000), with the extractive sector starting to contribute most of the gross domestic product and foreign exchange and accounting for most of the new job opportunities. In some countries, the growth of the manufacturing and services sectors has facilitated a shift from agriculture, reducing the pressure on land, thus slowing deforestation and, in some cases, even reversing the process. Dependence on land for the production of goods declines drastically as materials-based production improves or is replaced by knowledge-based industries. The attendant impact on forest cover is evident from the historical experience of most developed countries, and recently also from that of some developing countries.

Limits to structural shifts

If structural shifts in the economy can alter forest resource use patterns fundamentally, a key question is: What is the probability of such shifts taking place in most countries in the foreseeable future? Alternatively, what are the chances that such changes may not take place at all in some countries? In a highly interlinked and competitive global economy, the economic niches are increasingly being redefined and redrawn, largely on the basis of comparative advantages. This may limit significant structural changes in the economies. For example, development of the manufacturing sector in many developing countries remains constrained by low investments, small markets or the lack of competitiveness of local production. Manufacturing in developing countries is often agriculture-based and its expansion tends to have a negative effect on forests because of increased raw material requirements. In several developing countries, the indigenous manufacturing sector has relied on highly protected internal markets. Economic liberalization is, however, resulting in major changes, and removal of import restrictions is often undermining local industries because of increased competition from cheaper, and at times better-quality, imported goods.

Although a number of countries are hoping to expand their services sectors to enhance income and employment, there are inherent limitations to such an option. Most of the high value-added service sectors, including mature sectors such as banking, commerce and shipping as well as the newer sectors such as entertainment, information, technology and finance, are centred in the developed countries, whereas most developing countries focus on services at the low end of the value-added spectrum.[1] Even in the case of the "brain-power industries", which a number of countries are hoping to use for fast-track development, most of the services occupy low value-added niches - *e.g.* processing of credit card bills, "screwdriver technology" such as electronics assembly, and similar labour-intensive operations that take advantage of cheap labour. The potential for generating surplus from these activities is limited and there is very little scope for significant structural shifts.

Windows of opportunity provided by the development of the extractive sectors (*e.g.* oil and natural gas) have often been lost through private appropriation of the benefits or through the tendency to become too dependent on a single sector (Dale, 2000). When oil or natural gas (or for that matter any booming sector that generates a significant inflow of foreign exchange) becomes an important source of income, it pushes up the exchange rates, makes other sectors non-competitive and discourages diversification of the economy (a phenomenon sometimes called "Dutch disease" because it was first noted in the Netherlands in the 1960s following the discovery of large reserves of natural gas).

BEYOND STRUCTURAL CHANGES: SOME POSSIBLE FUTURE SCENARIOS

If major structural shifts are unlikely to take place in the foreseeable future, what will be the direction of changes and what are the implications for forestry? Will the future bring a situation of stable equilibrium that will enable most countries to bring about the necessary changes within the framework of the market mechanism? There are reasons to believe that changes could take place in new directions, largely because of the emergence of new fault lines in the global social and economic system.

Fault lines, disequilibrium and new scenarios

Predictable changes take place under equilibrium conditions in which it is easy to assess and understand the changes, whereas system-wide changes occur under disequilibrium conditions. As society develops and evolves, there are always new fault lines along which major changes take place. During most of the latter half of the twentieth century, the political-economic divide between the centrally planned and the market economies was the major fault line at the global level (Thurow, 1996). This divide tended to overshadow other fault lines. The collapse of the Soviet Union and the emerging dominance of the market mechanism were the major changes during the past decade.

However, the end of the cold war has brought to the surface some dormant or unnoticed fault lines. Many of the issues, such as authoritarianism, violation of human rights and corruption, that were overlooked (or sometimes tacitly approved or tolerated) to keep countries within a given political bloc have become unacceptable. Emphasis on democratic decision-making is percolating to the subnational level, resulting in efforts to devolve administrative responsibility to local levels. The emphasis on human rights is lending support to ethnic, linguistic and religious groups, often fuelling new conflicts and sometimes undermining the concept of nation states. Market-oriented development, especially increased flow of funds, technology and goods and services, is blurring national borders (Giddens, 1998). Alliances or interest groups that cut across countries are becoming important players in the economic, social and political fields. In a less polarized world with weaker nation states, new tensions are already surfacing which will have direct and indirect impacts on natural resources management, including forestry. These tensions include:

- The increasing economic and political power of transnational companies, the emergence of a large number of interest groups based on religion, ethnicity, language, etc., and consequent emerging conflicts;
- High material consumption and deteriorating environments;
- Widening inequalities and worsening distribution of income emerging from unequal access to skills, resources, technologies and markets.

Ruptures or tensions along the above fault lines are already evident. The large multinational companies, which are now consolidating their position through mergers, acquisitions and improved technologies, will face increasing pressure. In many cases, future conflicts will be between very active, highly networked organizations, representing the interests of a multitude of diverse groups, and economically powerful multinational corporations.[2] The rapid growth of information and communication technology is accelerating the process of change along the fault lines, with the increasing emergence of groups and communities that have a strong local base but interact in a transnational environment. There may also be strong local groups that are excluded from or unable to take advantage of the market mechanism. While the market mechanism has triumphed in recent years, it will have to adapt to a situation where new non-market systems are becoming an integral part of the economic and institutional framework.

PREPARING FORESTRY FOR CHANGE

Since there is uncertainty about the precise nature of future changes, it is difficult to indicate how forestry today should gear itself to adapt to those changes. Although in many situations the traditional problems will persist for some time, it is imperative that foresters and forestry organizations foresee changes beyond the immediate future and are prepared to adapt to emerging situations. Some of the most important changes that could be envisaged are outlined in the following sections.

Changing role of organizations

National-level objectives, programmes and plans will become less relevant as local communities, subnational entities and the private sector become the dominant forces in forest resources management decision-making. It is quite likely that forestry as a distinct sector or profession will cease to be relevant and will be integrated into a much broader framework encompassing all natural resources sectors. The role of national entities, especially hierarchically structured organizations such as forest departments, will diminish considerably. The ability of traditional forestry organizations to influence forest management - in terms of knowledge, resources, acceptance by stakeholders, etc. - has already declined considerably. Possible future roles for such organizations will be as impartial arbitrators to facilitate resolution of conflicts among the large number of emerging players, and as apex bodies to develop standards for the various practices that are acceptable to the diverse players. Since most of the productive functions will be performed by farmers, communities and private enterprises, the public sector is unlikely to have any major role in wood production. At most its domain will be limited to managing public goods and services, often under contractual arrangements.

New skills and approaches

Environmental issues, especially those relating to protection of watersheds and biodiversity, are becoming more prominent. Although ensuring the sustainability of public goods supply will still be a public domain responsibility, the approach will have to be very different from what it is today. Those who have to refrain from their usual activities to ensure the flow of environmental services for others will require appropriate compensation. Improved technologies for resource monitoring and fine-tuned management practices will have to be developed to ensure that all parties comply with their obligations. While the criteria and indicators for this will be developed at the national level, implementation and monitoring will take place at the local level. Substantial improvement may also be required in negotiation skills and conflict management at the local level, and highly transparent systems of resources management will have to be implemented.

RECENT AND FUTURE INNOVATIONS AND ADVANCES

Sawmilling and various machining processes

The most notable recent and expected advances in sawmilling and machining concern productivity gains obtained through mechanization, boosted by the introduction of computer-supported manufacturing processes. In traditional sawmilling, human judgement was used to obtain the best possible yields from the material. With modern techniques, cutting patterns are optimized on the basis of the qualitative and quantitative collection of data concerning inputs, which is made possible by the development of increasingly refined sensors of form, defects and irregularities. The data are then analysed in terms of the demands to be met, the state of stocks upstream and downstream of the work position, and priorities based on the product's rate of return. Products are produced following specific client orders in a relatively short delivery time, with stocks kept low. The development of these techniques allows considerable improvements in product quality and business competitiveness.

On-line control of sawmilling operations has recently been developed, often using visible, infrared or X-ray imaging techniques. For example, sawmilling machinery is increasingly equipped with sensors, such as lasers, that take account of the tool's behaviour or the blade's passage through the wood in order to assist and optimize manual control. With the recent arrival of new tools resulting from dynamic analysis and acoustic control techniques, the development of very economical sensors can be envisaged for the structural classification of woods, and even for the detection and pinpointing of certain defects or irregularities. Thus it is possible to foresee quality control based on classification by a variety of criteria such as size, the number of acceptable defects, colour, aesthetic properties and dryness. The extent to which these various criteria will be considered depends on the size of the company, the type of raw material and the market sector. Woods with a high level of irregularities (*e.g.* knots in softwoods) have already benefited from the application of this technology, which is becoming widespread in countries where the softwood industry is predominant. Processing of conventional tropical hardwoods can benefit only marginally from this technology because these woods are not subject to major defects such as knots. However, the technologies could be of great interest in tropical regions for the processing, grading and quality control of timber from plantations, such as eucalypts or tropical pines.

Further developments are expected mainly in the areas of data and information gathering on the milled product in both primary and secondary processing, and in the area of interactive quality control of the process through analysis of the main parameters, such as surface quality or precision of sawing. It will also be possible to reduce significantly the noise pollution produced by saws, while changes in these sounds can be used to determine the condition of the blades and possibly to guide cutting parameters such as feed speed. In surface machining (planing, moulding, etc.) there is a trend towards very high-speed machining, which can give a surface quality similar to that produced by sanding, while speeding up the work. Until recently, the rotational critical speed, related to the first mode of blade vibration, seemed to be an insurmountable obstacle for turning machinery, but rotation speeds between the first and second critical speeds can now be envisaged, thanks especially to advances in tensioning and computer-run saw-guides.

Complex multi-axial machining systems carrying out multiple simultaneous operations, such as the "work centres" available in the furniture industry, are gaining ground, as a result of developments in felling equipment, the widespread use of diamonds and the increased use of ceramics. These new cutting materials make it possible to adapt the tool's wear resistance to the high cutting speeds of such equipment and lengthen the use between sharpenings.

In veneering, automatic log-centring devices allow gains in time and, hence, in material. As in sawing, vibration or sound sensors will allow interactive control of veneering machines based on analysis of their performance. Furthermore, peripheral drive has made it possible to use very small-diameter logs and to reduce peeler cores to a few centimetres. On-line control systems, defect sensors, etc. also allow optimization of certain stages, such as wet clipping while the product and its various components are monitored during manufacture. The wear resistance of the steel used for blades and pressure bars will be improved by nitriding (the deposition of thin, extremely hard layers, possibly using plasma torches) and associated techniques. Research on jet cutting techniques (using lasers, water jets, etc.) will very probably remain marginal in the timber sector. The use of lasers is likely to be of interest only in the case of complex forms of machining (cutting rather than planing) or non-edge cutting in thicknesses of less than 40 mm. Conventional machining systems using chip removal methods are still by far the most cost-effective for all straight-edge cutting, and are likely to remain so.

Drying

At present, the most widely used drying technique is based on temperature and humidity control with forced ventilation. The low-temperature kiln dryer, which saw its hour of glory in the 1970s, has been almost abandoned as being too slow and entailing high risks of deterioration for fragile woods. High-temperature drying (at over 100° C), developed mainly for softwoods and some lighter hardwoods, tends to be unsuitable for solid hardwoods. The following technologies are now being developed for drying hardwood species that present difficulties and dry particularly slowly:

- Vacuum drying with superheated steam - the most effective of Vacuum technologies;
- High-frequency heating followed by a vacuum cycle - particularly recommended for thicker pieces.

Advances in modelling will allow the development of more effective and, above all, more reliable control methods. Continuous measurement of changes in the moisture level of wood, above the fibre saturation point, during the drying process is still a difficult technical problem. The absence of a reliable method of measuring high moisture content rates, other than weighing either the wood stacks or test samples, under industrial conditions considerably limits the effectiveness of the most recent generation of monitoring techniques, especially during the first steps of the drying process. Ultrasound and high frequency are among the avenues currently under research.

Wood treatment and preservation

The environmental impact of existing preservation processes has considerably hampered the development of new chemical approaches. The European directive on the use of biocides will further limit both the number of active substances used in preservation and their fields of application. Over the last decade, many pesticides have been removed from the market, and industrialized countries have banned some well-known families of products such as creosotes and pentachlorophenols. There is also much debate on aqueous-phase heavy-metal-based products of chrome or copper combined with arsenic or boron, which are used to improve durability. Since there is no real substitute for these products, a definitive ban could put an end to the outdoor use of the species in most common use today.

Solvents used as a medium for active materials have been developed considerably, and emulsion systems have been refined for non-water-soluble substances. In addition to the development and refinement of preservative products with a low environmental impact, many lines of research on the margins of preservation chemistry are being pursued and are likely to find alternative solutions. These include the following:

- The development of construction systems favouring "passive" preservation (*i.e.* systems that eliminate permanent contacts between water and wooden structures) and/or using naturally durable species, of which the majority are tropical;
- The use of natural insect-repellent or anti-appetant substances from tree species that have natural resistance to certain insects, particularly termites;
- The combination of durable with non-durable woods in new materials or products;
- Very high temperature treatment using hot air or a heat-conducting fluid, possibly followed by a treatment bath;
- Chemical processes such as acetylation to make wood hydrophobic, polyethylene glycol treatment and grafting of inert molecules on to the hydroxide bonds of cellulose;
- Resin impregnation procedures followed by accelerated polymerization and possibly preceded by thermo-plastification deformation cycles.

The most advanced and probably the most innovative lines of research concern the use of:

- Bacteria capable of destroying certain insects or fungi;
- Hormones that upset pest growth factors, especially insect ecdysis.

The identification of genes responsible for the natural durability of particular species could, in a more distant future, make it possible to modify the intrinsic durability of certain species through genetic modification. Such advances will depend to a large extent on the international debate on genetically modified organisms (GMOs).

Gluing and coating products

The development of chemical products has been affected by the need to take account of the effects of volatile organic compounds on human health and the environment. In the glue sector, the main innovations have concerned the development of low-formol products, in line with changing regulations. The sector now has an extremely wide range of specific products that meet both the technical demands of use conditions and the demands arising from issues of industrial implementation. Available products range from hot-melt to hot-hardening glues, with a whole gamut of intermediate products. New developments include glues for greenwood, glues for rough sawnwood with thick joints, and electromagnetic wave-accelerated polymerization technologies (high-frequency, microwave, etc.).

In finishing, problems arising from the use of organic solvents have been partially solved by the development of products with a high content of dry extracts or by the refinement of water-based systems in the paint and stain sector. Research is also being carried out on products such as radiation-polymerizable products and powdered products that do not use solvents. The lifetime of exterior facing systems has been improved by the addition of anti-ultraviolet (UV) components, improvements in the visco-elastic behaviour of films and slowing in the physicochemical degradation of existing products.

The main handicap in the exterior use of wood, however, is that the physical appearance still deteriorates too quickly. Research has shown the primary role played by the species selected in the lifetime of a finishing system; this lifetime can be doubled by the use of certain species. Research is focusing increasingly on this subject and investigating the performance of species-product combinations. Other work suggests that, in due course, it will be possible to coat wood based materials with metallic materials such as copper, aluminium and zinc, or with carbon-based "diamond" coating.

The main developments in the future could concern improvements in the productivity of finishing procedures through the development of single-coat products to replace the classic system of three successive coats. Regarding associated technologies, haze application systems that make it possible to obtain thin and very homogeneous coatings could come to the fore. Flat composites could see the development of powder finishing techniques. The use of electromagnetic wave devices could be expanded to speed up the drying of finishes in secondary processing industries, especially the furniture sector.

Composites and reconstituted wood

The past 30 years have seen the development of many types of board in response to two main demands on the part of the furniture and construction sectors - for thin flat materials and for large areas of board. Plywood was the first response to these requirements, but its manufacture required the use of high-quality wood (well-formed, cylindrical logs, straight trunks, etc.). Fibreboard and particle board produced from defibrated or fragmented wood do not require these qualities, and allow an optimal use of forest subproducts such as timber from coppicing or from thinnings, or residues from the processing of solid wood. Variations in the type of glue used, and in the size, shape and orientation of the particles, allow the production of board with a wide range of different properties to meet the technical and economic requirements of any application.

Research is likely to develop in two main directions:

- Prior physical and chemical treatment of particles to give them desirable properties such as regularity of size and natural durability;
- Chemical treatment to help particles stick to one another, allowing the manufacture of board without the addition of glue.

Plastic-wood composites should also see interesting developments. These allow the recovery of fine residues such as sawdust. Sawdust, when mixed with plastic in a proportion of up to 50 percent, produces materials that are mechanically more resistant than basic plastics, as well as being partially biodegradable. The normal techniques used in manufacturing reconstituted elements from offcuts or small-sized pieces continue to be developed, particularly in relatively unindustrialized countries, as alternatives to the use of waste in situations where fibre or particle composites are not a valid option.

WHAT FUTURE FOR THE TECHNOLOGICAL GAP BETWEEN NORTH AND SOUTH?

For most of the twentieth century, research and the resulting technological innovations were mainly in the hands of temperate industrialized countries. Seldom have innovations first developed specifically for tropical woods been transferred to temperate or cold-climate woods. One case that comes to mind is stellite tipping, a technique developed for metal machining and adapted by the International Cooperation Centre on Agrarian Research for Development's Forêt programme (CIRAD-Forêt), the French forest research and development organization for tropical regions. The technique has now been extended to all sawing operations and is used as widely for temperate as for tropical woods.

Tropical countries are increasingly in a position to benefit from the technological developments already introduced in the North, adapted to their needs if necessary. The emergence of a strong demand for wood as a construction material and a source of energy in certain tropical regions where there are highly industrialized wood sectors, such as Southeast Asia, has transformed the trade structures inherited from the twentieth century and will contribute to the increase of specific needs for technological innovation. The role of the emerging countries of the South will be vital in preventing the widening of the technological gap between the wood sectors of the South and the North.

THE TRANSITION TO PLANTATION FORESTRY

More than 4 000 years ago humans began to change the way in which they met their food needs by moving from foraging and hunting to crude cropping and herding and, finally, to modern agricultural cropping and livestock raising. Today this transition in agriculture is largely completed, except in a few parts of the world.

It was only in the latter half of the twentieth century that humans began to make similar changes in forestry: a transition from a primitive mode of gathering forest bounty, which is created solely by nature (old-growth harvesting) to the development of the science of wood production (silviculture) (see Figure). By the middle of the twenty-first century, the transition to tree cropping will be largely completed, and the greatest part of human wood consumption will come from planted forests, most of them intensively managed.

Through tree growing, commercial wood can become a crop, as in agriculture, to be planted, tended and harvested. Tree growing allows for a choice of location and species, as well as providing the opportunity to provide intensive management. It is only with tree planting that investments in tree improvement are justified. Improvements that result in higher yields or desired traits only make sense if those gains can be captured physically and in the market. Under an intensively managed regime, trees can be grown much faster with the desired traits. Climatic conditions for this are particularly favourable in parts of the subtropical and tropical world.

The trend towards tree planting will undoubtedly continue. It is driven by two powerful forces: economics and environmental concerns. Promising economic returns on tree planting have been realized in some locations for several decades, especially in advantageous areas of subtropical and tropical South America, Asia and Africa, where biological growth rates are high. In recent decades, large areas of previously unforested land have been converted to planted forests across the globe from the Nordic countries to New Zealand and from South Africa to South America. Although much of the activity has been in the industrial world, many of the most promising opportunities are in developing regions. Many planted forests are on land that was previously in low-productivity agricultural uses. The economic returns on planted forests, especially high-yielding intensively managed forests, are sufficient to continue to induce substantial investments in plantation forestry (Sedjo, 1999a). The general trend to high-yielding planted forests is receiving additional momentum from environmental concerns which have resulted in prohibitions on harvesting from some old-growth and secondary forests and regulations that make such harvesting more expensive. As the environmental movement continues to exert pressure for the protection and setting aside of more native and natural forest areas, less of this type of forest is available for logging and the costs of obtaining wood from these sources are rising.

New forest practice acts and codes are increasing harvesting costs in natural forests in many countries. For example, Kajanus and Karjalainen (1996) calculate that new forest policies in the Nordic countries, which emphasize the preservation of biodiversity, have increased harvesting costs. Similarly, it is estimated that the recently revised forest practices code in British Columbia, Canada, which deals with roading standards, riparian zones and harvesting practices, has increased the costs of timber harvests in the province by at least 15 percent (Haley, 1996). As harvesting costs have increased, harvests have declined in the coastal mountains of British Columbia.

A dramatic example of the reduction in harvests from national forests is the drop in the United States, from roughly 60 million cubic metres annually in the late 1980s to about 15 million cubic metres today. Pressures resulting from environmental concerns are likely to be amplified through the impacts of various efforts, such as those of the Forest Stewardship Council (FSC), to audit forest management practices or certify forest products. Although auditing applies to plantation forestry as well as natural forest management, it tends to raise the cost of natural forest management relative to plantation forestry. This is especially true where the planted forests are established on lands that were previously in other uses, *e.g.* agriculture.

The forces discouraging logging in natural forests are unlikely to go away in the foreseeable future. The pressures to reduce harvesting in old-growth and some second-growth natural forests will add to the attractiveness of the movement to invest in planted forests.

INDUSTRIAL WOOD DEMAND

Over the next several decades, the demand for industrial wood will undoubtedly increase, but not dramatically. Since the mid-1980s industrial wood demand has stagnated, with consumption of industrial wood remaining at roughly 1 500 million to 1 600 million cubic metres annually (FAO, 1984-2000). This levelling out can probably be attributed in part to the decline in production within the countries of the former Soviet Union as markets have replaced the earlier command and control economy. Increased recycling is probably another contributing factor. Since the decades immediately following the Second World War, the increase in wood consumption has been following a downwards trend. This trend towards stabilization has occurred in the face of substantial growth in the world's population and at a time when the global economy as a whole has been experiencing rapid economic growth, particularly in highly populated Asia.

Furthermore, if the latest United Nations (UN) population projections are to be believed, there is a good chance that by the middle of the twenty-first century the population of the planet will stabilize or even begin to decline (UN, 1998). In this context, it is difficult to see the basis for any dramatic acceleration in the rate of growth of industrial wood demand. By 2050, total world industrial wood demand will be higher than it is today, but not very much - perhaps 50 to 75 percent over 50 years (Sohngen, Mendelsohn and Sedjo, 1999).

SUPPLY

The predicted transition to planted forests would make the sources of wood supply dramatically different from what they are today. The author estimates that, at present, 22 percent of the world's timber harvest comes from what are essentially old-growth forests, while 34 percent originates in planted forests, and only 10 percent in fast-growing forests, which in 2000 consist of fast-growing exotics (Sedjo, 1999b).

Table. Current and forecast global harvests, by forest management situation

Forest management situatio harvest	**per cent of global industrial wood**	
	2000	**2050**
Old-growth	22	5
Second-growth, minimal management	14	10
Indigenous second-growth, managed	30	10
Industrial plantations, indigenous	24	25
Industrial plantations, fast-growing	10	50

By contrast, Table 1 forecasts that by 2050 up to 75 percent of industrial wood will come from planted forests, and about 50 percent from fast-growing forests. The fast-growing forests may no longer consist largely of exotics; rapid improvement is being made with certain temperate species, *e.g.* poplar, so that a large fraction of future fast-growing forests may be made up of improved domestic species. However, in much of the subtropical developing world, exotics are likely to dominate. Much of the planted industrial wood would come from high-yielding intensively managed forest operations. In some parts of the world, as is already the case in certain areas of the United States and Canada today, these operations may be called "fibre farms" and may operate on rotations as short as five to six years. However, most of the increased plantation output is likely to come from the newly created planted forests of the subtropics. Globally, fast-growing planted forests will encompass an area of perhaps 200 million hectares, or only about 6 to 7 percent of the world's currently forested area. Extensively managed natural forests that meet the auditing standards are likely to continue to provide a portion of the world's industrial wood, but largely for speciality products. Such changes have profound environmental implications.

The far more rapid growth associated with intensive management implies that huge volumes of wood will be produced from relatively small areas of land. Thus a trend towards planted forests and tree breeding does not imply, as some erroneously maintain, that vast areas of natural forests are to be replaced with planted forests. Accordingly most of the world's natural forests would remain for other purposes.Indeed, it has been seen (for example, in the National Forest System of the United States) that increased demands are placed on natural forests for other purposes as the production of industrial wood from these lands declines. Often this change is driven by policy and regulations, but such policies would not be feasible in a world where demand for industrial wood is far outstripping supply.

Technique	Increase in yield (%)
Orchard mix, open pollination, first-generation	8
Family block, best mothers	11
Mass pollination (control for both male and female)	21

ALTERNATIVES TO WOOD

One of the challenges to industrial forestry is the contention that the world's fibre needs could best be met by substituting wood with annual fibrous plants, such as hemp and bagasse. However, the following factors suggest that the prospects for other plant fibres as alternatives to wood are likely to be small over the next 50 years. First, using these types of fibre sources involves a number of economic and ecological problems. A major economic problem is that they have a specific peak harvesting period: the crop must then be stored and preserved until it is to be processed, and both of these functions incur cost. By contrast, timber can generally be harvested throughout the year, or at least throughout a larger portion of the year than an annual crop can, so the labour and capital equipment can be used year-round. In addition, wood tends to resist deterioration better than non-woody plants do.

From an environmental perspective, it is difficult to see how an annual crop, planted and harvested every year, can be more benign to the environment than a tree crop harvested only 20 years after planting. An annual crop generates 40 disturbances of the land over a 20-year period, compared with two disturbances for a tree crop. Furthermore, the energy required for annual planting and harvesting, and the greater amounts of other inputs such as fertilizer almost surely make such an approach environmentally inferior to tree plantation (Sedjo and Botkin, 1997).

NEW TECHNOLOGY: AN EFFECT ON COSTS

While the costs of harvesting natural forests are rising because of the effects of both availability and restrictions on current practices, the costs of intensively managed plantations tend to be falling because of new techniques, improved plants and the like. Costs can be lowered in forestry, as they have been in agriculture, through the introduction of appropriate technology. New technology can reduce unit costs by increasing the yield or the output per unit cost. Similarly, innovation can make possible the same production at reduced costs, again reducing the unit cost.

Lower costs are likely to have two effects. In the short term, return or profitability per unit will rise. Over a longer period, rising profitability is likely to increase production, thereby lowering prices to consumers and eroding some of the increased profitability. The short-term effect increases profitability, while the longer-term effect leads to lower prices for the consumer. In the context of forestry, where there is a high degree of substitutability between plantation and natural forest wood for most purposes, lowering the costs of plantation forestry also has the effect of providing a financial incentive for the industry to continue its shift away from higher-cost natural forestry.

TREE IMPROVEMENT AND BIOTECHNOLOGY

As already noted, tree improvement has few applications in the absence of planted forests. As planted forests become more common, the actual and potential applications of tree improvement using traditional techniques or biotechnology increase. Table 2 shows the extent to which various breeding approaches have been able to generate increased productivity. Biotechnology is expected to serve as a useful tool in breeding. Genetic modification will probably usually be applied to superior trees obtained through traditional techniques.

Traits that are of interest to forestry and that are likely to be enhanceable through genetic engineering include herbicide tolerance, flowering control, fibre and lignin content, insect tolerance, disease tolerance, wood density, growth, stem straightness, nutrient uptake, and cold, wet and drought tolerance. Genetic modification of some of these traits, *e.g.* herbicide tolerance, has already been well developed in agriculture. The introduction of genetically modified organisms (GMOs) into agriculture, and now forestry, is very controversial because of a host of real or perceived risks to health, safety and the environment. It is difficult to predict the extent to which this issue will ultimately be resolved. However, it should be recognized that, globally, biotechnology need not be an "all or nothing" proposition. As with nuclear power, it may well be that some nations will utilize the technology while others will not. Thus, industrial wood from genetically altered trees could become common in parts of the world while remaining largely absent elsewhere.

The financial incentives for the utilization of biotechnology in forestry appear to be strong. Short rotation periods, which are emphasized in high-yielding plantation forestry (where rotations are typically from six to 30 years), provide an inherent financial advantage. Furthermore, the use of biotechnology, which is currently under development, to improve fibre characteristics and/or reduce lignin extraction costs would reduce processing costs, thereby improving financial returns. To the extent that costs of establishment and/or processing can be reduced or yields can be increased without increasing costs, net benefits can be achieved.

A CONFOUNDING ISSUE

Perhaps the greatest unknown for the future of forestry is the phenomenon of global warming and its possible effects (IPCC, forthcoming). Warming could not only change the global distribution of forests (although probably not greatly by 2050), but could also have an influence on the extent of forest plantations. Should forests be used as major global carbon sinks to mitigate the build-up of carbon in the atmosphere, afforestation of large areas could be undertaken as part of the mitigation process. This activity could receive further impetus should bioenergy be used on a large scale to replace fossil fuels. Finally, wood materials sequester carbon even as they are utilized, and they have a lower energy (fossil fuel) input in their production than most other materials, *e.g.* steel, brick or concrete. These advantages could provide additional incentives for the expansion of planted industrial forests.

DRIVING FORCES AND OUTLOOK TO 2050

The model used here in predicting the outlook for sub-Saharan forests and forestry to 2050 is a dynamic model that relies to a large extent on developments in the rest of the world. First we assume that Africa will pass through three eras of political, economic and social growth between now and 2050, characterized first by political stabilization, then by technological growth and finally by environmental concerns. A second assumption is that the influence of socio-economic, political and technological factors will become more favourable to forestry in the future. We also recognize the possible impact of global political forces that are beyond the control of Africa.

The era to 2020 will see massive and uncontrolled deforestation in Africa because of poverty, as Africa will be faced with poorly performing economies and forests will be exploited for poverty alleviation. According to the United Nations (UN), the region's population growth rate will remain as high as 2 percent per year into 2020. FAO has reported that 16 African countries still face exceptional food emergencies resulting from civil strife, population displacement and droughts. With food crisis, investment in forestry development will remain rare, and many African countries will not be able to raise internal resources for environmental protection programmes. External resources for forestry, in the form of aid and grants, will become more scarce and may even disappear by 2020; it appears that the end of the cold war was also the end of generosity to Africa. Debt cancellation will be minimal and will have no impact on forest management. Development bank loans could be the only way of obtaining some of the limited international resources for forestry development from now until 2020.

At present, the lending rates of development banks are not sector-sensitive; infrastructure and conservation projects are currently subjected to the same lending rates in spite of the environmental benefits of conservation projects, which are weak competitors for financial resources. While many African countries (particularly those in turmoil) are indebted to development banks, the possibilities for lower rates in favour of less competitive sectors appear dim. Thus, funding for sustainable forest management will remain a major problem into the year 2020.

The acute shortage of funds for forestry will have a highly negative impact on the environment and wood supply. This will be most clearly manifested in Africa's inability to meet its domestic needs for industrial wood. While public sector funds from government annual budgets will continue to improve in the future, they will remain seriously inadequate to bridge the gaps that have arisen through long neglect of forests. Over the next 20 years, forestry as a commercial enterprise will remain relatively unattractive to the private sector because of difficult access to land, non-liberalization of timber trade, low wood prices resulting from government price distortions and high borrowing rates.

Era of technological growth (2020 to 2040)

This will be a period of real development for sub-Saharan Africa, with advances in science and technology supporting agriculture and industrialization. Stability and good governance will begin to be seen in most parts of Africa. Food security will have been ensured in many countries, with significant development observed in terms of human resources and welfare, technology and industrialization. With industrial growth there will be less dependence on agriculture for employment.

Crop productivity per hectare will have improved drastically, enabling the agricultural sector to release some lands for urban and forestry uses. Private foreign capital will probably have started flowing to Africa because governments will have privatized a larger part of their energy, water, communication and transportation sectors. From 2020, a more stable population can be expected, as well as a more open economy with a high level of trade liberalization and improved infrastructure. Regional integration for improved intra-Africa trade will be strengthened in all subregions of sub-Saharan Africa. Taxation is likely to become more effective with improvements in revenue collection systems. Interest on borrowed capital will have fallen to a single digit. In this era, however, African natural forests will be highly degraded as a result of long development neglect and exploitation of the resources for economic growth. We expect environmental degradation resulting from forest loss to reach a peak by 2040, with flooding and erosion posing the greatest development challenge in the region. Today's national parks and game reserves will perhaps remain as the only natural forests to survive. Well-kept films may be the only records of the herds of zebras and antelopes that were once abundant in Kenyan and Tanzanian parks. Much biodiversity will have been lost because of the dryness of many African ecosystems. Wood import bills for some countries will be equal to today's oil import bills. Thus, in terms of development, those countries of Sahelian Africa that lack mineral resources will begin to fall behind the countries of moist Africa, as a result of the burden of energy and wood bills.

With significant wood shortages leading to higher prices and increased trade liberalization, private entrepreneurs and farm families will begin to grow more wood on abandoned farmlands with the encouragement of tax incentives, which governments at all levels will begin to use to stimulate conservation and timber growing. Forestry research will start to receive the support of the private sector and greater appreciation from governments. Governments will also begin to promote conservation, which will become an election issue. By 2040, communities will become less relevant as agents for forest protection, and private ownership of land will become dominant. Government forestry institutions will have fewer staff, and their role will centre more on formulating policy and initiating environmental legislation. The focus of forestry training will shift from extension to providing skills for forest land management as an economic enterprise.

Era of growing environmental concerns (2040 to 2050)

While, by 2040, the structure of African forests will have changed through the loss of many indigenous species and the introduction of exotic species, the years 2040 to 2050 will see a rapidly growing concern for the environment. With great improvement in incomes, forests will begin to serve more recreation and watershed protection needs.

Africans will become less dependent on forests for their energy needs as electricity and gas become more abundant. These developments will be very positive for the environment and conservation. Small-diameter logs will have greater importance for the supply of construction timber. Forest fires will be brought under control as a result of private investments in timber growing. More international agreements will have come into effective operation, and forests will no longer be national assets but of strong regional concern. Unfortunately, this may happen only when the resource has already gone. We therefore call on governments, international organizations, non-governmental organizations (NGOs) and individuals to double their efforts to avert this situation.

ASIA AND THE PACIFIC

A Russian proverb says that when we predict the future, the devil laughs. So what if the devil laughs? Let me predict the future of forestry in the Asia and the Pacific region in the year 2050, 50 years hence, and see who has the last laugh, the devil or us foresters! By 2050, forestry will have become the victim of the social, economic, scientific and political changes that will take place in the region, and in the world, over the next five decades. The walls of communism will crumble everywhere, including in China, and capitalism will be the norm of economic activity. This, together with the spread of globalization into all countries, will have great economic impacts on forestry. The development of science through research and development will have progressed by leaps and bounds, and forestry will benefit significantly, especially from biotechnology.

Privatization of forestry will become the norm in most countries. The public sector, *i.e.* the forestry departments, will be responsible solely for enforcement of the law and the collection of revenue. The private sector will be repsonsible for all other functions of forestry, including the management of forest plantations, natural forests and parks. This will ensure the practice of more integrated forestry in which upstream activities, traditionally in the hands of the public sector, will be linked directly to downstream activities, which traditionally have been in the hands of the private sector. This evolution will have far-reaching impacts on the development and practice of forestry as well as on the forestry profession in Asia and the Pacific. The natural forests of the Asia and the Pacific region will be diminished in area. Tropical forests will be limited to Kalimantan and Irian Jaya in Indonesia and small pockets in Malaysia and Papua New Guinea. The Philippines and Thailand will have little natural tropical forest left. Subtropical forests will be limited to isolated reserves in Southeast Asia, Australia and the Indian subcontinent, while temperate forests will still exist in small reserves in the temperate regions of China, the Indian subcontinent, Australia and New Zealand. However, the use of all these remaining natural forests will be limited to conservation, production of water and recreation, while all wood fibre needs will be met by forest plantations. Foresters will have finally recognized that they must manage the whole ecosystem, including fauna and flora, and must not be limited to managing only trees. These areas will be managed by the private sector and the public good will be one of their most economically valuable products.

FOREST MANAGEMENT OBJECTIVES

Forest recreation will be the commodity in greatest demand, and ecotourism will have been developed to a true science. Ecotours will be so popular that forests will have to be zoned for different intensities or types of eco-activities, and specific recreational management plans will have to be developed to manage recreation within the remaining natural forests. In many countries the supply of freshwater will be a major problem, and the natural forests will also function as water catchments. These forests will be managed for the production of water which will be sold to public utilities.

The greatest forest-related income will come from bioprospecting permits and agreements. The remaining natural tropical forests of Asia and the Pacific will be opened to international bioprospecting, which will become a global industry equivalent in importance to prospecting for oil. Products that will be sought include pharmaceuticals, cosmetics and other natural products for the food industry such as natural colourings, new herbs and condiments for food. Regulations will be in place to control these bioprospecting activities through a protocol under the United Nations Convention on Biological Diversity and national legislation to ensure equitable benefits and sharing of technology.

Tropical countries in particular, where the greatest biodiversity exists, will be proactive in promoting these activities. Adequate control measures will be in place to ensure that the owners of the biodiversity also receive the benefits from these activities. Bioprospecting will be such a lucrative business that maintaining the forest for such purposes will benefit forest owners more than timber harvesting. AIDS and a host of other deadly diseases will no longer exist, because cures will have been found through such bioprospecting arrangements. All the region's wood fibre needs will be produced by forest plantations, including rubber and oil-palm plantations, which will be recognized internationally as forest plantations. These forest plantations will be managed on short rotations for the production of wood fibre which will then be processed into a variety of value-added reconstituted products. New and sophisticated processing systems using wood fibre as input will provide finished products to order. The whole production process will be automatic and controlled by computers.

Carbon sequestration will be one of the most important functions of forests, as new protocols approved under the revised Framework Convention on Climate Change will allow and promote carbon credits and trading of carbon in the international futures commodity markets; biodiversity credits and transpiration credits will also be traded in the international markets. The global demand for freshwater will exceed supply, and the trading of water credits as a commodity will be on the horizon. Certification of timber and forests will be the norm; no wood fibre or timber will be sold either locally or internationally without certification for sustainable management. A system for managing the chain of custody will be in place and practised widely. In addition to the Forest Stewardship Council (FSC), other certification bodies from countries in the South will be recognized and accepted by the international market.

THE FORESTRY PROFESSION

With the management of forests under the private sector, consultancy companies will mushroom in many countries of the South, giving new life to forestry and invigorating the profession. Forestry will have become a respected profession again, even greater in stature than medicine or information technology and engineering, mainly as a result of the demand for the outdoor life and sustainable lifestyles. The forestry profession will be governed, not only by national rules and regulations with their codes of conduct, but also by an international global forestry council under the auspices of the UN. Forestry schools will proliferate as the demand for forestry training intensifies. The Asia Pacific Association of Forestry Research Institutions (APAFRI), formed in 1995, will be the premier forestry society in the region and its mandate will have been expanded beyond research, to cover the professional practice of forestry.

SCIENTIFIC ADVANCES

Biotechnology in forestry will bloom, in terms of both institutional research and the application of research results. Planted trees will be genetically modified to meet certain needs, such as maximizing carbon sequestration, fibre production or any of a wide range of other modifiable traits. Numerous transgenic trees with many different characteristics, such as resistance to diseases and maximized growth rates, will be developed and grown in plantations - although forestry will still be a long way from producing the "ideal" tree. Growth rates of 100 m3 per hectare per year will be normal in tropical species. In 2050, a new transgenic tree that grows fast, absorbs carbon efficiently and produces edible shoots and fruits will be undergoing genetic modification to produce sap that can be used as fuel to drive automobiles without any processing. Scientists will aim to start economic production of this fuel by the year 2075. Because it will be possible to produce large volumes of wood fibre, biomass energy will be the preferred energy source for power generation, and numerous small biomass power generation plants will cover the rural tropical regions of the Asia and the Pacific region.

Nevertheless, through the development of forest plantation technology, the traditional high-value indigenous timbers of the region will also be grown in selected areas through enrichment planting of degraded forests or in forest plantations. These tropical species will be grown on a 20-year rotation, but their wood will be expensive and will cater for international niche markets. The arid and semi-arid zones of the region, for example in China, Mongolia, Central Asia and Australia, will benefit from the development of new transgenic plants and trees that can survive and grow under low moisture conditions. The semi-deserts of the past will be green with plants and trees that can withstand extreme low-moisture conditions. Some of these trees will have been genetically modified to produce strong and long adventitious roots that can seek moisture deep in the soil, and aerial roots that are able to absorb moisture from the air. The aerial roots, stems and branches will also have nodules that absorb nitrogen from the air. When they have been greened, these large expanses of land, once the eyesore of the earth, will generate enough moisture in the air through transpiration to encourage the formation of clouds and even rain. The greening of the deserts, one of the oldest human ambitions, will slowly become a reality. Economical technologies for the desalinization of seawater will allow irrigation of semi-arid and arid regions close to the sea, which will become greened for habitation and agricultural production.

Progress in the greening of the deserts will help overcome the major problem of population in China and India, which will still be the countries with the largest populations in the world. Agricultural production will increase sufficiently to meet the domestic needs of these countries for basic foods. With the land available, the technologies developed and the human resources available, the Asia and the Pacific region will have become the leading producer of food and wood in the world.

EUROPE

Fifty years is less than a complete forest rotation in most parts of Europe, so forest managers today must take actions with a mental picture of conditions 50 to 100 years in the future. Projecting present trends into the distant future is also one of the best ways of understanding the present. Speculations of this nature, however, need to concentrate more on social and economic trends than on technical forestry matters, as the former have the strongest influence on forestry practice.

It is assumed here that Europe will remain peaceful and prosperous, possibly even more prosperous than it is at the beginning of the twenty-first century. If Europe were again to be subject to war or a major catastrophe, natural or human-induced (such as a nuclear explosion), then forests, as well as people, would be devastated and the first priority of foresters would become the protection or reconstruction of what remained. Such events are impossible to foresee, but they cannot be excluded entirely.

Before looking forward, it is wise to look back. What have been the main structural changes influencing European forestry over the past 50 years? Since the Second World War, perhaps the most fundamental changes have been a huge rise in general prosperity and the current transition from centrally planned to social market economies.

In the forest sector, with the exception of the years immediately following the Second World War, fellings have stayed well below increment and forest area has expanded steadily. There have been few major changes in silvicultural theory and practice: the trend towards intensive monocultures favoured in the 1950s and 1960s has been reversed in response to criticism from an increasingly well informed and environmentally sensitive public. Silviculture has returned to earlier principles, which are more cautious and less economic. Forest work has become less difficult and dangerous. Improvements in chainsaw design and increased mechanization have improved working conditions but have also made wood harvesting in many parts of Europe somewhat capital-intensive.

The balance between the main roundwood assortments has changed radically: fuelwood has become insignificant in many areas as the prices of non-renewable energy have reached historically low levels (in real terms), and logs are increasingly losing ground to pulpwood.

The rise of products based on reconstituted wood has provided outlets for almost all parts of the tree. Very little raw material is wasted. Recovery of waste paper is now standard all over Europe. Recovery of used wood is starting to follow the same trend. Since the 1970s, the population of Europe (especially northwestern Europe) has become increasingly aware of environmental issues and, with growing prosperity, increasingly unwilling to accept the justification of environmental damage on economic grounds.

THE EXTERNAL ENVIRONMENT

Energy prices will be significantly higher in 2050 than they are today, and there will be systematic encouragement of renewable energies, including wood. Oil will lose its preponderance as an energy source by 2030. No single energy source will replace it, but increasingly sources will be adapted to the various uses. Many political and administrative functions will be internationalized or decentralized to a regional and local level.

The nation state, although still important, will no longer be the unique focus of political power. In many areas, European Union (EU) decisions will be crucial. The EU will extend to the frontiers of the Commonwealth of Independent States (CIS) and will have a major standardizing influence in all areas of life, including the distribution of wealth between richer and poorer parts of Europe. It will also be a more democratic organization than it is today. However, on the global level, the influence of the EU and other large powers (the United States and Japan) will be counterbalanced by a rise in the influence of other regional powers.

The Russian Federation will pass through further periods of extreme tension, even chaos, but will emerge as a major regional power with a reasonably competitive economy. Decline in population (by 20 to 30 percent) will be a significant problem, and hundreds of millions of hectares will be left as wilderness (except for mineral extraction and forestry).

Multinational companies and international NGOs will have great influence, economically and on public opinion. Structural change will be impossible without at least the tacit support of both groups. Stable or declining population levels, combined with economic prosperity, will make labour in Europe even more expensive (in relative terms) than it is now and will encourage mechanization and automation in every field. New technologies in the communication field and elsewhere will continue to reduce costs and increase effectiveness. There will be an enormous building boom between 2000 and 2030 in the countries of Central and Eastern Europe.

DEMAND FOR GOODS AND SERVICES OF THE FOREST

Wood-based products will continue to hold a significant part of the construction and furniture markets, provided that producers continue to innovate, market their products vigorously, keep prices competitive and maintain a favourable environmental image. Composite products with tailor-made specifications, manufactured in large, capital-intensive units from homogeneous, low-quality raw materials will dominate. Traditional sawnwood with its low processing input, irregular characteristics and high raw-material quality requirements will become a luxury or niche product. Only a few companies and regions of production will have the scale of operations, capital reserves and expertise to be serious global players in this competitive environment. In Europe, such players might include the Nordic and Baltic countries (increasingly seen as a single region); northern Spain, Portugal and southwestern France; Ireland and Scotland; Austria; Poland; and the northwestern area of the Russian Federation. These areas will be marked by very intensive silviculture and a high concentration of processing plants. Elsewhere markets for wood as raw material will be weak, and profitability (if there is any at all) low.

The demand for recreational facilities will dominate in all forests near centres of population and tourism destinations. Conflicts between user groups will intensify, with forest owners (public or private) having to arbitrate. Consumers will be increasingly sophisticated and demanding as regards both product performance and environmental aspects. For example, European consumers will no longer accept wood from natural forests - but this will be of marginal importance as all natural forests will have been permanently protected or converted to semi-natural (*i.e.* managed) or plantation forest, and plantation-grown timber will be much cheaper.

Energy will be the major new market as, under the influence of climate disasters attributed to greenhouse gases, governments finally use the price weapon to discourage the use of non-renewable energy sources3 and encourage wood supply and demand through vigorous actions. There will be many small wood-burning installations; wood will be almost the only energy source for heating in rural areas and a significant one elsewhere. Wood-based fuels such as ethanol or methanol will be manufactured on a large scale to replace some uses of oil.

FOREST POLICY AND MANAGEMENT

Strict standards of biodiversity conservation (protection of key habitats, wildlife corridors, etc.) will be enforced everywhere, without exception. In the intensive wood production areas, which will be owned by either forest industries or large private owners linked to the industries through contract or share ownership, specialized wood production companies will carry out most management tasks. The whole system - from plantation to harvesting, transport, processing and marketing - will be highly optimized, and all links in the chain will be in constant contact with one another. Genetic improvements and intensive silvicultural regimes will bring much improved yields and shorter rotations. Trees will be selected genetically, not only for fast growth, but also for wood characteristics such as reduced lignin for pulping. In most cases, the end use of the tree will be known when it is planted. In these areas, forest ownership and management will be profitable and there will be an active market in forest land.

Outside the intensive wood production regions, the general utility of forests to society will be recognized. Forest owners' income flows will come from wood sales, mostly to local wood energy markets (which will be closely regulated, as the markets for electricity, gas or public transport are today); they will also receive public payments to compensate costs not covered by wood sales, in exchange for significant restrictions on their freedom of choice. User fees will be seen as unfair and difficult to administer, as road use fees are today. For a very few forests with high recreation value, entrance fees (similar to those already made for motorway access in some countries) will be charged.

In areas with high recreational use, the public authorities may have to take over the day-to-day management of the forest, as small private owners will not have the requisite skills. However, in more remote rural areas outside the intensive wood production regions, most forests will hardly be managed at all, reverting slowly to a more natural state and expanding naturally on to former agricultural land. A large part of forest managers' time will be spent running public participation exercises: the profession of forester will be seen as a "people" job, not a technical one.

Climate change will modify site conditions and influence yields. Some areas will become more competitive (*e.g.* southern Finland and Sweden), while others will experience acute problems (*e.g.* Mediterranean forests threatened by fire and desertification), but overall the sector will adapt successfully. Europe will have a few "Kyoto forests" (forests established and managed specifically as carbon sinks and taken into account in the auditing process of the Framework Convention on Climate Change), mostly financed by arrangements with local or national power companies, but most of the activity linked to carbon sinks will take place in areas with better growing conditions and cheaper land.

FOREST SECTOR ISSUES

Certification will no longer be a contentious issue; either certified products will have a small niche market or (more likely) all forest products will be routinely certified. At the local level, conflicts among user groups will be the most important preoccupation. At the national and EU levels, discussion will centre on the public funding that can be made available for forest management in non-intensive wood production areas. At the international level, political discussion will focus on problems of trade policy (a "level playing field" for wood producers). It will prove difficult to distinguish payments for the management of non-intensive forests (which will be considered legitimate) from subsidies to intensive wood production areas (which will be considered, in theory if not in practice, illegitimate). As today, the main preoccupation of most forest owners will be how to cover their costs and make a profit, while satisfying all the needs of society.

LATIN AMERICA AND THE CARIBBEAN

The Latin America and the Caribbean region still has vast land areas covered by forests, but the potential represented by these forests has not yet been fully recognized and developed in most countries. What will the scenario be in 2050? This question can only be answered by first looking back. As in most parts of the world, forests in Latin America and the Caribbean were in the past considered to be an obstacle to development with relatively low economic importance. For a long time, the region had a negative balance of international trade in forest products.

In the early twentieth century, even Brazil, currently the region's main producer and exporter of forest products, was a large importer of timber. For many years the United States, Canada, Finland and Sweden exported large volumes of lumber to Brazil. Factors arising from the First and Second World Wars helped reverse this situation, but in Brazil, as well as in other countries of the region, the consolidation of the forest sector did not start until much later. Land, agricultural and forest policies developed by some countries during the 1960s were perhaps the most important elements leading to changed perspectives related to forests and forestry in the region.

Forestry policies had a substantial impact, particularly in Brazil and Chile. In the 1960s these countries developed a fiscal incentive programme to support the establishment of forest plantations. The plantations, mainly based on pines and eucalyptus, soon made available uniform and low-priced raw material, recognized as an important element in attracting the capital needed to develop the forestry industry. Land and agricultural policies led to the occupation of tropical forest areas. As a result, large volumes of high-quality and low-priced logs were made available. During the same period, the tropical timber industry flourished in Asia. Tropical timber products gained new markets, and thus opened new perspectives for investments, particularly in the Amazon basin. The regional economic crisis of the 1980s, globalization and environmental pressures were key elements in introducing recent changes to forestry developments in the region. Chile was able to modernize and open its economy faster than other countries in the region, attracting capital and developing the potential represented by the established forest plantations. Brazil faced a long period of high inflation and economic stagnation. Lacking capital and being less attractive to international investors, its forest sector developed more slowly. In any case, these two countries were the only ones in the region to develop the forest sector (at least to some extent), mostly based on plantations.

Brazil and Chile served as models to other countries. Lessons learned from the fiscal incentive programmes in these countries were used to develop fiscal and other incentive mechanisms further. Incentives are now important instruments for the expansion of forest plantations in Argentina, Uruguay and Paraguay, and other countries have shown willingness to adopt similar models. The low sustainability of agricultural projects in the Amazon, and environmental pressures were important factors for the development of new forest policies. The process started in Brazil, where several legal instruments for further regulation of forestry activities were put in place. The process is now spreading all over the region. Based on an ample discussion, Bolivia adopted a new forestry law in 1996. Peru approved a new law early in 2000, and other regulatory mechanisms are under development.

The model adopted can vary from country to country. In Brazil, for example, the model is based on privately owned production forests, while in Peru and Bolivia forests are the property of the state and are made available to the private sector as concessions. All models, however, have incorporated the principles of sustainable forest management.

REGULATION - A KEY DETERMINANT OF THE FUTURE

Regardless of the model adopted or the type of forest, all forestry-related activities in most countries of the region are heavily regulated. In many cases decisions taken by governments are influenced by international pressure. As a general rule, regulatory measures have been increasing in forestry, while in other sectors policies for deregulation predominate in response to the globalization process. Apparently, most governments of the region have failed to put in place the proper mechanisms to make environmental concerns and development policies compatible. Low managerial capability is the main problem, and this is not likely to be solved in the next few years.

In spite of the efforts made, international cooperation has not been able to help solve this and other existing limitations. In retrospect it seems that international cooperation has been an expensive and inefficient mechanism. Governments are motivated internally and externally to increase regulations. This requires new structures to enforce the legal instruments, and little attention has been paid to the efficiency of the process. The process is augmenting, and will continue to augment, costs to governments, and is transferring costs to the private sector. Some countries have promoted the decentralization of public administration, involving state and municipal governments in forestry, environmental monitoring and other related issues. This has been considered a way of enhancing the involvement of stakeholders and of facilitating the adjustment of regulatory and development instruments to local specific conditions. The principle is correct, but the results in most cases have not been positive. Decentralization has generally resulted in overlapping of regulations, multiplication of regulatory bodies, more conflicts and additional costs. As a result of these developments, companies and countries of the region are becoming less competitive, and lower competitiveness creates more limitations for the adoption of sustainable forest management, the final goal. Other global market players are less regulated and have a competitive advantage.

AN IMPORTANT ROLE FOR PLANTATIONS

The regulatory trend will continue in the coming years and, for some countries of the region, it will probably take at least 20 years to reverse it. By that time forest products based on native sources are likely to have lost most of their market. Plantation forest areas in Latin America and the Caribbean will continue to expand. In most countries of the region, plantation forests are heavily regulated at present, but they will be less regulated in the future. In addition, from a purely economic point of view, plantations are more productive than native forests. These factors will make plantation-based products more competitive. Technology will further help plantation forests, making it possible to produce higher-quality and more uniform material more quickly. In the future, better and cheaper wood products will be produced from plantation-grown wood.

Not all countries of the region will benefit from forest plantations. Plantations are a long-term investment, and not all countries are able to ensure the legal, political and economic stability required by investors. In addition, countries with limited local markets and poor infrastructure will have less of a chance to develop forestry in the coming years. Current developments related to forest plantations clearly indicate that in the near future the Southern Cone of Latin America will be among the most important forest-product producing regions of the world. Pinus and Eucalyptus plantations will be the main source of raw material. Eucalyptus species will gain market from tropical timbers.

Capital will continue to flow into the Southern Cone countries, mostly via private investors. Large corporations will gradually replace the existing industry. Forest ownership and industrial production will be highly concentrated. The regional market will grow faster than the global market and will be important for local producers, but the region will also become an important player in the international market. Native forests in the future will be managed mainly for environmental purposes. Native forest production areas will be reduced, as native forests are set aside for environmental reasons or because of lack of competitiveness.

More and more money from international cooperation and other financing mechanisms (carbon sequestration, debt swapping, biodiversity protection, etc.) will be invested in the protection of native forests. Several governments of the region will accept these funds and will pass and enforce laws and regulations that will continuously reduce production from native forests. This will be an easier and faster means both of solving immediate problems related to lack of capital and of achieving the expectations of the countries' populations. Governments will thus reduce social pressure and gain political stability. Future generations will judge the success of this decision.

NEAR EAST

Peace and Environmentally Friendly Governance

It is hoped that, well before 2050, regional conflicts, intergovernmental disputes and internal strife will have been amicably resolved. Coups d'état will give way to democratic processes. The only way of assuming power will be through the polls. Commitment to the environment and to sustainable development will have high priority in party manifestos and election pledges. Promises will be kept and commitments effected. With better governance, public opinion will be more influential, and the roles of civil society, NGOs, the private sector and local communities will be enhanced. All of these are likely to have a positive influence on directing more attention, efforts and resources towards the environment, particularly tree planting. They are also likely to promote greater efficiency in the use of resources for this objective. Budgetary allocations for armament, war, security and the like will be shifted to more humane and constructive uses. The bulk of national budgets and bilateral and multinational funding will be directed to reconstruction, sustainable development, environmental rehabilitation and human welfare.

THE NEW FORESTER

With the settlement of internal strife and the attainment of social stability, urbanization and automation, employment opportunities will be needed. What better source of opportunity than forestry and tree planting? Discharged soldiers and draftees in national services will be retrained and deployed, together with their earthmoving equipment, trucks, water tankers, etc., to work in reconstruction and environmental rehabilitation. They, together with forest guards and foresters in general, will be trained to forego the police mentality and to accept social fences in place of barbed wire ones. This transformation will entail radical changes in the curricula of forestry schools in order to give more emphasis to non-traditional services and products rendered by forests, especially the protection of watersheds and watercourses, desertification control and NWFPs. Greater attention will also need to be given to social forestry, to the removal of "Berlin Walls" between forestry, agriculture and horticulture and to implanting contemporary concepts such as national forestry programmes, biodiversity and sustainable forest management. The ultimate objective is to produce the awaited "new forester".

In addition to discharged soldiers, other social sectors besides those currently involved in forestry and tree planting will become involved, particularly women. Indicators of this are already evident in some countries such as the Sudan, where increasing numbers of female students are joining faculties and departments of forestry (as well as other areas of higher education). Indeed, a recent batch of forestry graduates in the Sudan was made up of 11 women and only one man. Many women, as individuals or as groups, are already forest owners or are involved in private, homestead and community forestry. If the trend continues, by 2050 it may be not the women, but the men in forestry who are seeking equal opportunity.

Sustainable forest management and conservation

As attention shifts and environmental awareness grows, factors detrimental to the environment, forestry and trees will be reversed. Decades of the practice of sustainable development will finally have brought the concept home. Sustainable agricultural production will be sufficient to meet national and regional demand, or nearly so. Urban sprawl will not encroach on arable land but will inhabit deserts and mountainsides. Environmental concern will not only stop the curtailment of forest and woodlands, but will make more land available for them. With better living standards and the availability of alternative energy sources, particularly electricity, kerosene and butane gas, the need to use fuelwood or to collect it as an income-generating activity will be diminished. So will communal grazing and transhumant livestock rearing have decreased. Instead, the forest estate and arboured areas will be expanding.

With increased awareness of forests and woodlands as common property in which all people have an interest, deliberate destruction and arson on these lands will decline. Conflicts over land use in forested areas and disputes with foresters will no longer be protested by the belligerent setting of fires - currently a serious problem in the region - as the need for forest protection gains greater recognition. Sand dunes may continue to encroach on human settlements and property, but they will be undergoing stabilization and fixation. Wildlife sanctuaries, recreational parks, green spaces, botanic gardens and arboreta will be in the process of establishment everywhere. Farms, homesteads, civic centres, roads, canals and railways will be lined by multiform, colourful and seasonally flowering trees.

The resolution of internal and interboundary conflicts will facilitate the allocation of resources and ease access for the sustainable management of common watersheds, the establishment of common shelterbelts and the harvesting of Run-off from rainstorms. One of the important activities in watershed management will be tree planting and conservation. This will eventually render more and better-quality water for all purposes, including irrigated forests, green areas, shelterbelts and scattered trees. Similarly, the resources once spent on war and on reversing the consequences of war will be freed up for allocation to desalinization of seawater and recycling of excess water from irrigated fields, sewage effluents and industrial wastewater, and the use of this water for irrigating trees. The benefit from the extra water made available will be maximized through the development and adoption of improved irrigation systems. Of course, even with the resulting improvement in moisture regimes, the increased numbers of trees and shrubs in the landscape and the reduced setbacks from overgrazing and factors of vegetation removal, it is not possible to expect that the environment will revert to what it was in prehistoric times. Still, many species of flora and fauna could probably be restored, either from buried seed banks or through deliberate reintroduction, to the levels of the not-so-distant past.

NORTH AMERICA

Increased Appreciation of the Environmental Value of Forests

Perhaps the most pronounced change in North American forest management over the next several decades will be a continuation of the dramatic shift in public perception concerning the value and appropriate uses of forests. In particular, publicly owned natural forests will become increasingly valued for the environmental services they provide - notably watershed protection, biodiversity conservation and carbon sequestration - instead of just their wood and other forest product values.

Clean, reliable water is and will surely remain one of the most important products of forests throughout North America, essential not only for irrigated agriculture in the western United States and northern Mexico, but also for industrial and residential use throughout the continent. In fact, the National Forests in the United States were created in large part to reverse the deterioration of watersheds during the nineteenth century and restore them to health, a process that occupied much of the first half of the twentieth century. Today, these forests encompass some 3 400 watersheds which provide drinking-water for more than 60 million people. As the economy of Mexico continues to diversify and expand, thousands of expanding municipalities will depend on water from forested lands for both domestic and industrial use. Forest-based recreation, already a high priority in much of North America, is likely to become even more important as per capita productivity rises and leisure time increases in all countries, and as the urban middle class in Mexico continues to grow. As species continue to be lost worldwide, forests will become increasingly valued as reservoirs of biodiversity. In North America, attention will likely continue to centre on the species-rich forests of southern Mexico and Mesoamerica. But biodiversity will also be an important factor in temperate forest management, as witnessed by current efforts to modify forest management practices along the west coast of the United States and Canada in order to protect wild salmon populations.

Finally, as the effects of global climate change become more apparent, the role of forests as both carbon sinks and moderators of climatic disturbance (such as flooding) will take on a new meaning. Implications for management objectives include not only the addition of carbon sequestration to the goals of multipurpose forest management on public lands, but also specific reforestation and forest protection projects on private lands, in response to incentives provided by carbon markets.

In order to increase the role of environmental services in forest management, at least three fundamental changes in North American forest management practices will be necessary. First, substantial cost and effort will be required to restore North American forests to ecological health. This is particularly true in the western United States, where a combination of extensive harvesting and fire prevention, however well intentioned, has unfortunately led to undesirable changes in species composition, stand structure and fuel loads, leaving many forests vulnerable to uncharacteristically intense fires and disease spread. Similar, although perhaps less severe, challenges face Canada and Mexico.

Second, innovative mechanisms that reflect adequately the full value of environmental services will need to be adopted in public policy-making and market structures. For example, healthy, functioning watersheds save local communities throughout North America billions of dollars in water filtration costs. Yet the environmental services of forests have traditionally been treated as a "free good," with inadequate recognition of their true value or the costs associated with maintaining them. This is beginning to change; New York City, for example, recently decided to invest US$1 500 million in watershed management and reforestation as an alternative to paying up to US$8 000 million for new water treatment plants. To correct the problem of undervalued environmental services, it is likely that such market-based practices as conservation easements, carbon trading and "true cost" pricing of water, recreation and hydroelectric power will become more common.

Third, increased attention will need to be given to the social dimensions of forest management. Recreational use in National Forests in the United States has grown from fewer than 20 million person-days in 1950 to approaching 1 000 million person-days today, yet insufficient effort has been given to understanding the nature and management implications of this dramatic shift in forest use. As managers adapt to changes in public attitudes towards and uses of forests, research will be needed to clarify social priorities and improve understanding of human interactions with forests. As appreciation of the importance of social and institutional arrangements grows among forest managers, it is likely that new approaches will be adopted to ensure transparency and public involvement in decision-making. Criteria and indicators of sustainability at both the national (*e.g.* the Montreal Process) and the management unit (*e.g.* certification) levels, the devolution of decision-making to local institutions, and innovative approaches to public-private partnerships are just a few examples of possible means of improving transparency and public involvement in forest management. Finally, to manage forests on a landscape scale, it is likely that new mechanisms will be developed for voluntary coordination of land management across ownership boundaries, including international borders. The presence of Mexican and Canadian firefighters in the United States during the severe fires of 2000, as well as similar assistance to both countries from the United States in recent years, is testimony to the potential of such cooperation.

WOOD AND FIBRE PRODUCTION

North American forests will not, however, cease to be important sources of timber, fibre and other commercial products. On the contrary, it is probable that North American wood and fibre production will increase in the next few decades in response to overall domestic and international demand, which is likely to grow, despite product substitution effects. However, this production will probably by concentrated increasingly in privately owned plantation forests that are specifically dedicated to fibre production, rather than in publicly owned, natural forests. One of the reasons for this trend (which is already much in evidence) to continue is the reduction of harvesting on publicly held lands because of the depletion of commercially available stocks and public concerns regarding the compatibility of logging and environmental services. The introduction of new fast-growing hybrid trees adapted to a wider range of growing environments will both help to increase the comparative economic advantage of plantation-grown wood and extend its range to new areas.

This does not mean that commercial harvesting will necessarily disappear from natural forests during the coming decades. Carefully regulated harvesting is likely to be an important tool for managing forests for multiple benefits (including restoring them to health) for decades to come. In fact, a priority will be to find new means of utilizing small-diameter wood as part of the process of restoring ecological health to Western forests with high fuel loads. In addition, small, family-owned forest plots throughout the region, particularly in the northeast and the tropics, are likely to continue to be managed for high-value wood products, among other objectives. It seems probable that genetically modified organisms (GMOs) will have a revolutionary role in fibre production in the coming decade and beyond. Given the legitimate concerns over the wisdom of creating and deploying GMOs, especially in wildland habitats, it seems likely that their contribution to fibre production will be focused on tree plantations and agricultural crops, combined with new processing technologies for composite materials. Perhaps even more influential will be the impact of GMOs on forests via the agricultural sector, through the concentration of crop production and subsequent reforestation of marginal lands, the extension of modified crops on to lands previously unsuitable for agriculture, or both. Finally, it is impossible to overlook the potential of GMOs either to harm forests through the introduction of novel invasive species or to be of benefit to them through the reintroduction of such species as the American chestnut and the American elm.

8

Clones in Nature and Forestry

Clones are common in Nature, and they have been used as a tool of domestication for very long time and are used in large-scale forest operations. It seems to be important to remember that to put clonal forestry into perspective, The earliest land plants lacked vascular tissue and had inefficient systems of sexual reproduction and dispersal and clonal reproduction was the typical way of propagation in nature. Clonal growth was advantageous enabling physical dominance of large areas through horizontal growth.

Clonal propagation is still important in nature for many species. Many of the commonest clones are herbaceous but the habit is also found in woody plants throughout the world. In temperate broad-leaved woodlands, root suckering is a common form of clonal growth and there are good examples in the genera Populus and Prunus. Coppice is a natural growth form that has been exploited for centuries in European forests. Betula, Carpinus, Corylus, Quercus, Salix and Tilia are all genera with the ability to self-coppice.

The use of vegetative propagation of trees as a tool in their domestication has a long history. An early use was to propagate good clones of fruit trees and this has been done for thousands of years, thus humanity has long experiences of cloning trees.

For specific cases clonal forestry is often regarded as a "new technique", and "biotech" leads associations to the public to something more radical than most cases of clonal forestry. To put things into perspective it has to be remembered that many of the techniques used in modern "seedling forestry" are also "new", and the problems and uncertainties with that is often larger than the uncertainties connected to clones. Some cases where clonal forestry has been much used are described, *viz.* Sugi, Eucalypts and poplars.

SUGI

The first known use of vegetative propagation for forestry purposes was in Japan with Sugi (Cryptomeria japonica D. Don, a conifer) in the 15th century. Since then reforestation with monoclones or clonal mixtures has been widely used in Japan (see Ohba 1993). Cryptomeria japonica is the most widely cultivated tree species in Japan comprising nearly half of the ten million ha of Japanese plantations. Almost half of the plants used are clones (praxis differ in different part of Japan and with different owners).

The extreme uniformity of the plantations has probably contributed to greater than average damage from typhoons and heavy snow. Attacks from Sugi bark borer and bark midge have increased in severity as the area of Sugi plantation has grown, but there is no evidence of altered disease or insect virulence related to clonal forestry, despite careful monitoring.

Clonal forestry is most common on the south-western island of Kyushu where about 100 cultivars are in use. Many foresters use only 1-3 cultivars of local origin in their plantations so that a single clone may cover as much as 2 hectares.

Allergy to Sugi pollen is a health problem in Japan as a whole, but is only a minor problem on Kyushu with its large use of clones. Cloning by cuttings for several generations has reduced the incidence of male flowers, so these plantations produce little pollen. Analogously one may speculate in that a reversion of North European birch forestry to clonal forestry in a very remote future may reduce allergic reactions with birch pollen.

Eucalypts

Clonal forestry is more common with broadleaf species than with conifers. Probably the most successful example is the eucalypt plantations in countries in tropical and subtropical climates like South America, South Africa and Portugal. The total area of eucalypt plantations today is about 30 million hectares and about half of this area is planted with cloned material.

The eucalypt plantations in Brazil cover an area of approximately 3 million hectares. A succession of eucalypt species has been grown in Brazil due primarily to disease problems. Clonal forestry started in the late 1960:s and is today the dominant form of eucalypt forestry in the country. Progeny from an Eucalyptus grandis female (male sterile) growing in an arboretum were found to be natural hybrids which were resistant to pests and diseases. The offspring of this tree are believed to be Eucalyptus grandis ? Europhylla hybrids. This knowledge was rapidly adopted by the forest industry companies and they selected clones within the open-pollinated offspring from this single tree, which means that a small number of clones originating from the same family are widely grown over large areas in Brazil. This has and is still changing with most companies having good breeding Programmes backing their clonal Programmes. The clones are deployed in huge monoclonal blocks. This situation seems favourable for a pathogen, but nothing has happened yet - despite the possible risks of a disease in a monoculture. Today the companies have become more risk-conscious and are developing breeding and clonal testing Programmes that include strategies for clonal diversity in space and time. A typical Programme is now deploying about 20 clones each year and these clones will be replaced with time as new clones are selected. The production landscape forms a genetic mosaic. New genetic material has been imported to widen the genetic base for breeding.

Poplars and willows in Europe

Vegetative propagation of poplars (Populus sp.) has a long history in Europe. Organised clonal forestry started in the beginning of the 20th century. Monoclonal plantations of poplars have become a common land use especially on river plains in southern European countries like Italy, Spain and France. Some of the best poplar clones in use were selected as early as during the first half of the 20th century. Clonal forestry with poplars is common in countries with subtropical and temperate climates like Italy, Spain, France, Belgium, USA and Canada.

Single clones of poplars have been propagated extensively. There have been disease problems and the overuse of single clones and use of large monoclonal plantations has been questioned (cf *e.g.* Stelzer and Goldfarb 1997 p444). Poplar has covered over 100000 ha in one country and a single clone may comprise one third of this area. The focus has been to develop disease resistance within individual clones, but these clones have proved to be highly susceptible to new varieties of disease. The disease problem seems to be growing and has caused Germany to more or less abandon poplars and other countries may follow.

Nevertheless these incidents have not provided poplar growers with sufficient incitement to focus on more diverse alternatives. It is unlikely that monoclonal cultures and the wide use of some clones are the only explanation to increased disease, but they act as contributing factors. Legal and commercial reasons often favour the use of single clones (*e.g.* easier to use breeders right –UPOV - for single clones than varieties, commercial demands to know that varieties are up to specifications). Even willows and short rotation forestry in Sweden are threatened by disease. Resistance is broken down within a single rotation (it is decades in between replanting even when harvesting is done at intervals of a few years). Two evaluations of the Swedish willow breeding Programme have rather strongly recommended an increased number of selections to be made (more than a single selection released per year). Nevertheless, the growers´ organisations still do not encourage clone mixtures.

HYBRID ASPEN IN FINLAND

A Finnish forest industry company has recently started a clonal forestry Programme with hybrid aspen (Populus tremula crossed with Populus tremuloides) for plantations in Finland and Estonia. So far a few hundred hectares have been planted with micro propagated selected clones. A recent Finnish doctor thesis has appeared on the subject of hybrid aspen (Yu 2001). This raises the question if something similar could be done with birch. It may be noted, however, that Yu (2001) suggested that micro propagation costs are high and that the technique needs development.

OVEROPTIMISTIC STATISTICS

It is a clear tendency that statistics reported about the use of vegetative propagation, at least there it is regarded as modern hitech biotech, gives the impression of more rapid progress than actually occurs. *E.g.* Talbert et al. (1993, p. 148-149) reported, based on a survey, that four Swedish operators produces 7.3 millions rooted Norway spruce cuttings annually. Lindgren et al. (1990, p. 8) estimated the annual production to 5 millions and believed in a raise to 10 millions. However, looking backwards (Sonesson et al. 2001, p. 15), less than 20 million were planted in total till 2000 and now only some hundred thousands are planted annually.

Crop uniformity

As genetic diversity and possible uniformity is an evident and possible important aspect of clonal forestry. For evaluation of possible ecological impact diversity seems central. I will thus start to discuss disadvantages with a low genetic diversity in a stand. In other respects this paper focuses on the advantages of clonal forestry.

DISADVANTAGES OF A UNIFORM CROP

There are reasons to believe that the biological production can be higher in a genetically diverse crop than in a uniform:

- A single genotype demands the same things at the same time, and thus utilises the site worse than a mixture of genotypes.
- In a mix, another genotype may take over the ecological space left by a failed genotype.
- A disease spreads faster in a uniform crop.

These statements are supported by the average of a very large number of agricultural mixing experiments and a few experiments and experiences with forestry. On average, mixes performed some percent better and were less susceptible and more stable than the average of its components, but usually the production of the mix was not better than its best component grown pure. Even if mixes seem more productive on an average, there are many examples and experiments, when no advantage was found and some experiments where diversity actually seems negative for production. I conclude that from a production point of view an advantage with some degree of diversity is expected, but that does not imply that production is higher the higher the diversity is. But the superiority in biological production is not expected to be large and it is not certain for individual cases, but just a rough generalisation. This general statement ought to apply to birch also, thus biological production is assumed to be lower and more uncertain in a uniform crop, but the effect is probably small and practically unimportant (a few percent of biological production).

Advantages of a uniform crop

There must be considerable advantages in uniform monoclonal blocks as they are used frequently. Although the advantages with diversity are repeated and reiterated and very politically correct, it is seldom applied in intensive agriculture or husbandry. The main reason is that maximizing biological production and maximizing economical production is not the same thing. It makes it easier to manage the crop if it behaves uniformly. For a forester it may not matter very much if individual birches drop their leaves on different days, but a farmer wants the crop to become ripe the same day, as the whole crop is harvested the same day. Diversity is a usually a problem for the customer, who wants a uniform raw material for further processing. For administrative processes, like describing the crop, it is an advantage if it is uniform and does not change from specification over time and space. These problems are evident for many types of agricultural crops, and therefore these crops are usually uniform even when biological production would be higher with a diverse crop.

The dream of many foresters is to find the best genotype and use only that. This has been successfully done for other domesticated species. The uniformity itself in a clonal plantation has a considerable advantage from a management point of view. It is easier in the nursery to raise plants of a single genotype. It is easier to manage and harvest a uniform forest. It has advantages to handle a single type of material. A customer or processor of a product knows more about a lot if it is a specified clone than if just the species is known. A batch of uniform logs is worth more and easier to handle than the same amount of very variable logs. Even if interest is only in quantitative production and that is lower in a uniform crop, still the higher genetic performance of the best clones compared to the best seedlings available usually much more than compensates for the production loss by a uniform crop. Thus uniform clonal crops produce more than diverse seedling crops.

My opinion is that genetic diversity often is overemphasised in comparison with the higher genetic gain achievable at the cost of a reduction in genetic diversity for future forest production plantations in many countries (like Sweden) both for propagation with seedlings and cuttings. To get genetic diversity many clones are used in seed orchards and clonal mixtures and there constraints of relatedness of clones may be seen as desirable. This was not a disadvantage at an early stage of forest tree improvement, as it does not influence gain much if many phenotypically selected plus trees are placed in seed orchards, and in the early stages of cutting propagation there were physical constraints on the number of ramets which can be produced by each clone, so it is necessary to have many clones. But now this constraint often means selection is not intensive and the best genotypes are not exploited to the extent, which would be possible. However, clonal forestry is unlikely to deal with a single clone only. That does not seem a likely scenario for birch. Many clones will be tested and test results as well as experiences will accumulate of the best clones, so the opinion of what is the best clone will change over time. It is also dependent on the locality, the desires of the user and the availability. Even if uniform plantations are planted, there will usually be a mosaic of clones on the landscape level.

From the risk management point of view a clonal deployment philosophy offers advantages. If a clone is sensitive to a pest or pathogen the problem can be identified and managed. If silvicultural means to deal with a problem are insufficient, the stand can be salvage harvested and replanted. The clone which turns out inferior can quickly be taken out of production. It is an advantage for an operator which uses monoclone blocks to use several clones in the same time that it will become more evident what the strong and weak point of the clones are. A disadvantage with mosaics is that the uniform lots of wood from a stand are often too small to be practical as a specific variety at the end user, and actually the end user may get trouble with larger between lot differences if clonal forestry is used. Alternatively the clones can be grown in intimate mixtures, in that they most of the possible advantages of the diversity of seedlings may be gained by a mix of rather few clones.

When we are talking quality birch - and not quantity birch - the reasons for monoclonal plantings become stronger. When high quality of the product is concerned, it often becomes more important to have uniform quality also, while in a low quality product - as firewood - the quantity is usually relatively more important and a uniform and predictable product is less important. Attention on quality instead of quantity also generally coincides with a higher level of intensity, and at a higher level of intensity, clonal forestry often becomes more interesting. Thus considering the aim the project arranging this symposium has in mind; uniform clonal plantings becomes relatively more important than for forestry in general.

Improved clones can be preferable to seedlings!

There are a number of arguments for forestry with clones. It takes less time to multiply a genotype as a clone than to build a seed production unit with that clone as one of the parents and wait till that works. The time lag is actually much longer than till the first seed production, because a seed production unit is usually designed to last for many years. The risk for getting the seed storage empty means that forestry often want to have five to ten years foreseen seed need in storage. That is expensive, many seeds may never be used and the seeds actually used are less improved than possible.

A seed production unit is designed to meet a certain need of plants with certain characteristics many years ahead. If too many seeds are produced it means that the investment was unnecessary large. If too few seeds are produced it may mean that foresters must use genetically inferior unimproved seeds. If the propagation is done fast by vegetative propagation by *e.g.* micro propagated clones or somatic embryogenesis, it takes only a year to multiply a desirable genotype in the desired number of copies. Thus it is fairly easy to adapt to the latest news from breeders and customers.

There are large annual variations in seed production both in the forest and in seed orchards. In a seed orchard the reproductive output is variable among genotypes and the reproductive phenology may make clones incompatible, so the actual genetic set up of the progeny is a bit uncertain. These uncertainties can be avoided by clonal forestry. Many of our forest trees - like birch - can both spread pollen and produce seeds. That means that selfing can occur and the occurrence of selfing in a plant material will decrease the forest production. With the use of clonal forestry selfing can be completely eliminated.

In a wind pollinated seed orchard wild uncontrolled and unimproved pollen may contribute to pollination, clonal forestry is a way to eliminate this problem. However, for birch, seed orchards are small enough to be managed in plastic green houses, where the inflow of unimproved pollen is low, so for birch it is not fair to see this pollen contamination problem as an argument against open pollinated seed orchards.

In a seed orchard it is important to avoid related clones, as relatives get inbred progeny when they mate, which means that forest production drops because of inbreeding depression. Mating between relatives may be a worse problem than selfing in future seed orchards, as self-zygotes usually die; self-seedlings are often culled before planting; or early out-competed in forest, while milder inbreeding is more directly transferred to production loss. If clones are related in a clonal forest, that does not cause inbreeding depression, and thus there is little harm if related clones occur in a stand (actually this is very natural, trees in a natural stands are often related). The clones in the breeding population will be increasingly more related as breeding goes on, it will be easier to handle the consequences of this for the production forest with clonal forestry.

A seed orchard is a big investment intended to meet the seed need for a long period for a specific species used in a specified area, thus a considerable local market is needed to justify the investment. Conifer seed orchards are usually designed for a need of many millions plants per year over some decades. Multiplication of a clone can be seen as a smaller investment during a shorter period, thus the local market is allowed to be much smaller with clonal forestry, *e.g.* micro propagation for curly birch in Finland serves a market of 200 000 seedlings a year. A seed orchard has a stiff structure, its composition can be only slightly modified once it is established and it is not easy to justify several orchards heading for different characteristics for the same area. Clonal propagation is more flexible; an individual customer can choose the clones, which fits the particular needs for the customer and occasion. Sometimes a combination of rather rare characters is desired, a clone can be found which combines several desirable characters in a much more efficient way than dealing with seed orchards. It is often stated that clonal forestry reduces genetic diversity.

This is a generalisation, which need not always be true. Actually, clonal forestry provides a tool to choose genetic diversity at will, while when a seedling material is used, the genetic diversity is essentially left to chance. Probably those using intensively managed birch plantation for producing high quality timber with special characteristics would choose the option to reduce or eliminate genetic diversity on the stand level, because they found it economically sound to do that. It seems likely this is how the tool will be used. But clones in a clonal mix can be chosen so that the genetic diversity is larger than in a seedling material. Clones can be mixed with seedlings, that may be a good strategy not only for diversity, but economically in cases where cloned plants are much better, but also much more expensive. If the clones succeed well, their share at the final harvest will be larger than at establishment. Over time and space different clones will be used even in a system with monoclone cultures (like Eucalypts in Brazil). In such a system the genetic diversity among stands is likely to be much than in nature. Sometimes clones have been used for natural conservation in situation when good seeds could not be obtained. Clonal forestry may strengthen the trend to produce the needed wood at a limited acreage of intensively managed production forests, allowing larger areas to be set aside for other purposes. Thus the question how clones may affect diversity has many aspects and it is an oversimplification to state that it must be severely reduced.

A common situation is that some good seeds exists or can be obtained, but quantitatively insufficient for the plant need. If such a situation occurs unexpected and unforeseen, vegetative propagation may be the solution. A more common situation is that a few superior seeds can be produced by controlled crosses, but too few to meet the plant demand. Clonal propagation is often a way to make forestry with controlled crosses practical feasible. Clonal propagation may be seen as a way to amplify the impact of a few good parents or a tested cross by mating the parents and when multiply the seeds.

The genetic value of a genotype depends on the breeding value of its parents, but also how well the genes of the parents fit together. This may be called dominance. The later reason for genetic differences cannot be exploited in an ordinary seed orchard, but it can be exploited with clonal forestry. Hybrids sometimes result in good seeds. It may be difficult to make many of those seeds by controlled crosses, but hybrid seeds (or trees) can be multiplied by vegetative propagation. This is used in the Eucalypts and Aspen Programmes mentioned above. Personally I am not that convinced that hybrid superiority is so common as it is often stated. When hybrids are better than the parents it may be because the parents were not well adapted for the conditions where they were tested or that the hybrid combines characters of the parents for the environment and purpose of the forester in a better way than tested alternatives. Hybrid maize is a best seller, but the reason is more that farmers are unable to produce hybrid seeds them self, but has to buy them, and when it is good economy of a seed company to invest in hybrids, so they can market the seeds.

To get a genetically superior forest, it is desirable to use the genes of the very best clones. In a seed orchard it has to be rather many parents to avoid selfing and problems with lack of overlap in flowering time. This means a drop in breeding value. For clonal forestry the very best clone can be used. This is true even if in the complete absence of dominance. As dominance is seldom the major cause of genetic variation for the most important characters in forest tree breeding, this more effective exploitation of breeding value is usually a more important reason for clonal forestry than the exploitation of dominance variation, at least as long as we do not talk about hybrids of different species or races. Quantitatively the low importance for the superiority of cloning of a limited amount of dominance variation is demonstrated by *e.g.* Lstiburek (2000).

We have also special features for clonal selection. We can intentionally select clones, which do not waist resources on sexual activities but spend resources on stem wood production. Such clones are desirable for forestry. In a seed orchard it is not desirable with clones, which do not produce seed and pollen, so such clones would not be chosen for a seed orchard, and if they were, they would be unable to transmit their genes to the seed crop. Cloning methods have different characteristics. They are often very depending on the state of the material which is cloned, cuttings usually do not work well on physiologically old material, for somatic embryogenesis other tissue than immature seeds are often a difficult starting material. With micro propagation it is usually possible to multiply old trees or clones, and this is often an advantage.

Cloning and the cloning method can have effects on the physiological status of the tree. *E.g.* conifer grafts have not thick bark at the bottom of the trunk. The protocol for making clones can have effects on the plants. These effects sometimes look desirable and can be exploited. For example Norway spruce cuttings seems to suffer a lower mortality from pine weevil damage than seedlings (Mattson and Thorsén 1992). Clones usually origin from older and more mature tissues than seedlings and this affects the characters of the plants and may have effects also on the mature trees. I guess these differences to seedlings are not dangerous and nursery raised seedlings can also be claimed to be unnatural, and they my be different because of different growing regimes. I suggest that we can intentionally interfere with the characters of the plants by the production system and that clonal forestry gives additional options to use this in a positive way. Many of the factors I have mentioned above will contribute to that clonal forestry gives a genetically more improved forest than comparable seedlings. An example of the development of gain over time for the clonal option versus the seed orchard option is shown in Figure 1. This is based on a simulation of the Swedish Norway spruce (Picea abies K.) long term breeding Programme with inputs chosen to correspond to the real Programme as close as possible. Clonal forestry and seed orchards are different ways of harvesting gain from the breeding population and clonal forestry means that the genetic improvement reaches a certain level three decades before the same level can be achieved by seed orchards.

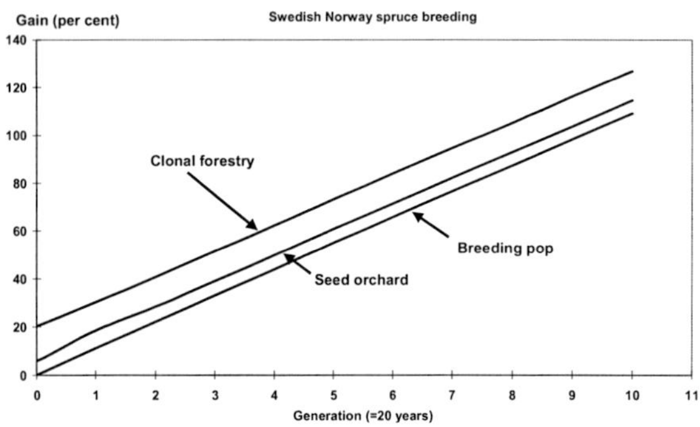

A comparison of the genetic value of three populations as a function of time based on a simulation of the Swedish Norway spruce breeding Programme. Genetic progress is primarily created in the breeding population. The production population is created by creaming the best available material from the breeding population either by clonal forestry or seed orchards. The gain unit can roughly be interpreted as percent gain. The details of the figure depend on exact specifications, but it gives an idea of magnitudes based on best possible estimates. Clonal forestry gives a higher gain than seed orchards; the difference at a given time is in the magnitude of 15%. The same gain is achievable from seed orchards; it just takes a few decades longer to get it. Modified from Rosvall, Lindgren and Mullin 1998.

More effective breeding with clones!

In the section above it was pointed out that clonal forestry might be a faster way to better forests than seed orchards. Good genotypes are multiplied as grafts in most current forest seed orchards, thus clones are useful also for producing genetically improved seedlings. But there is another - still more powerful - way of using the clonal tool to improve the genetic quality of forests. Clone testing has the potential to make the long-term breeding considerable more efficient. Even if clones are too expensive for direct use in forestry, their use in testing of genetic materials may help forestry to get better seeds in the future. A combination of testing clones for forestry and for long-term breeding has synergistic effects, thus such a Programme is more efficient than a Programme focusing only on one of these two aspects.

Higher organisms are diploid, thus have two set of homologous chromosomes and carry two packages of genes. The parents only transfer one package (a haploid gamete) to the progeny, and the composition of that gamete is different for each progeny. Thus, if offspring is used to test parents, only half of the genes in the offspring are relevant for the parent, and these genes will be different for each offspring individual. In contrast, in clonal testing all genes in the genotype are tested and no genes generate random noise in the statistical sense. The performance of a clone depends on its father, its mother and how the genes of the mother and father match together, thus dominance effects. However, dominance variation does not seem large enough for most characters of practical interest to be important (*e.g.* Rosvall 1999, Lstiburek 2000). Thus the performance of a clone in a clone test says much about what will be inherited to its progeny.

The accuracy by which the breeding value of a clone is determined increases with the number of ramets. The uncertainty caused by the environment decreases then environment is repeated. Thus clonal testing is a way of increasing the accuracy. One can see it as a way of increasing the heritability to decrease the environmental variance by letting the environment be the average of a number of individual environments, one for each ramet. The gain obtained by a genetic selection depends on the selection intensity and the accuracy of the test. Under the constraint of a fixed plant number, a lower number of genotypes can be tested if genotypes are replicated by cloning, and thus the selection intensity will decrease It thus becomes an optimization problem. However, if the individual heritability is not very high and the resource budget is not extremely tight, the optimum is to clone in at least some copies.

A seedling material is genetically diverse; the best genotypes grown pure are much better than the average of a seedling material. The performance of a seedling is mainly depending on its environment and not so much of its genes. If it is cloned, the average over the clone is more depending on the genes. Testing by clones can be seen as a way to eliminate the statistical noise. The heritability of clonal values will be much higher than the heritability of seedling values.

Each seedling is genetically unique; a scientific principle is to use replications when testing. Thus a seedling is a rather unsatisfactory material for genetic testing and conclusions based on seedlings will be dependent on model assumptions, which are uncertain and affected by statistical noise. Clone testing is less depending on model assumptions and a more direct measure and thus offers theoretical advantages!

In a series of investigations (*e.g.* Rosvall 1999; Danusevicius and Lindgren 2002; assuming conditions similar to the Swedish Norway spruce Programme), it has turned out favorable to test the breeding value of the candidates for the breeding population by clonal testing instead of phenotypic testing or progeny testing, if that is at all possible.

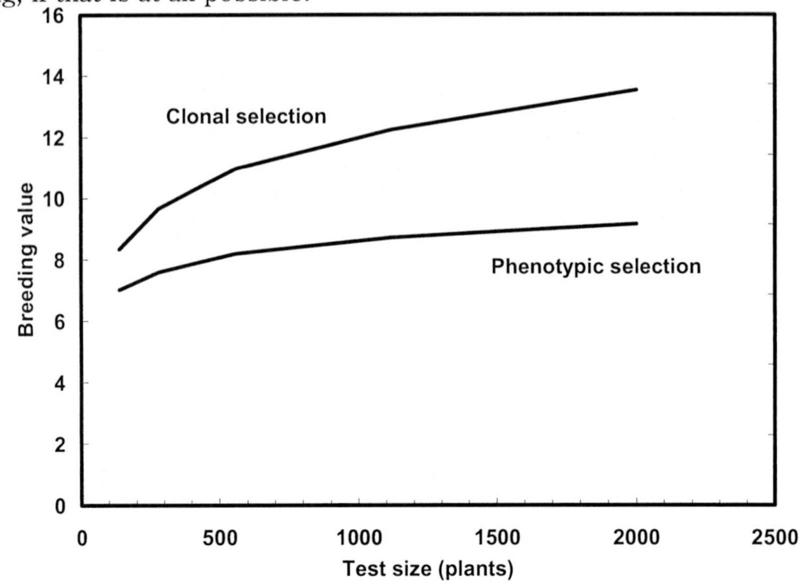

Comparison of clonal or seedling based testing for the Swedish Norway spruce long term breeding Programme (the X-axes gives the resource level, the test size for the real Programme is around 560 plants/parent). Clonal testing adds around 30% to the gain of the breeding Programme, the advantage is larger the more resources that are available for each parent (Rosvall 1999).

The superiority of clonal testing over phenotypic testing is demonstrated in Figure 2 (from Rosvall 1999 Figure 3). It is based on the same type of simulations of the Swedish Norway spruce long term breeding Programme as the previous figure, although it varies the resources available (thus the size of the tests). The resources available today correspond to 560 test plants per parent. The more resources available, the more reason to use clone testing to get a more accurate selection (compared to increasing the number of test genotypes for a more intense selection).

Lstiburek (2000) and Danusevicius and Lindgren (2002) studied the conditions under which clonal testing is a favourable alternative. The results show that clone test is superior to phenotype test and also progeny test under a wide range of conditions. Thus even if the genotypes or ramets are expensive;

and over a wide range of available resources; and in the presence of typical dominance variation; clonal testing can be recommended. In cases where the individual heritability is close to 1, clonal testing is less favourable. When the phenotype itself gives sufficient information about the genotype, the accuracy need not be improved by cloning. Such cases are rare for growth characters, but may be more common for some quality characters.

Numerous studies have shown that environment can cause lasting changes in phenotype that can be passed from one generation to the next, much as genes are transmitted (Schwaegerle, McIntyre and Swingley 2000). Such affects may limit the usefulness of clonal testing for breeding value, but similar objections could be made against progeny testing, and the effects must anyway be larger than usually observed (*e.g.* Lindgren and Wei 1994) to really matter. Natural genotypes are not designed for perfect cooperation; the Darwinian selection in the struggle for life unavoidable gives some preference to strong competitors. A certain degree of selfishness is an unavoidable consequence of natural selection. By clonal selection we ought to be able to identify and utilise the best collaborators, who act in the interest of the stand and not only the individual tree. Here is a case when Man may have the means to do better than Nature and identify collaborative clones.

A forest material is used over varying conditions in time and space and the forester wants a reasonable stable material over these variable conditions. Clones can be tested over time and space. That make breeders able to select clones, which perform well everywhere and anytime. Natural selection works just here and now. Again a case where Man may find more effective ways than Nature has used. Tree Breeders first select based on field tests and when mate the selections. Artificial crosses are difficult to organise in the remote places where field tests are localised. But if clones are tested, some copies could be placed where it is convenient to make artificial mating, like in archives outside the institute.

Foresters do not want trees to waist resources on sex, but to do breeding it is necessary to get reproduction to work. But clones, which have little reproductive output in the field, can have copies at the research station, which are manipulated to make mating possible. Thus with clones different individuals can be used for test and reproduction. Genetically modified trees will play a role in practical forestry and tree breeding at some time in the future. Such implementation is much easier, if clonal forestry is an established praxis.

EVOLUTION OF TECHNOLOGY FOR CLONING EUCALYPTUS IN LARGE SCALE

Since the first developmental stage until its recognition as an operational method of reproducing superior trees, cloning of Eucalyptus species is undergoing a continuous process of improvement through incorporation of new technical concepts and technologies at different phases of the process. Campinhos and Ikemori, 1983; Zobel and Ikemori, 1983; Ferreira and Santos, 1997; Denison and Keitzka, 1993.

In the last two decades, cloning Eucalyptus spp. has produced relevant progress for the forest companies, especially solving problems associated with diseases such as canker (Cryphonectria cubensis) and productivity improvement (Campinhos and Ikemori, 1983; Ferreira and Santos, 1997). Actually, the focus of cloning has shifted to industrial requirements, rather than remain limited to just disease resistance and volume increase. The wood properties that positively influence industrial processes and product quality are considered, especially where cloning has an important role to play.

In the era of global markets, the development of forestry plantations for industrial purposes must aim for, besides other objectives, increasing industrial competitiveness in the distinct market segments they interact with. In such scenario forestry based companies must consider the mode in which the forestry raw material can affect their competitive capacity. The modern concept of competitiveness includes producing products to meet the customer's requirements at low costs, in a sustainable manner and with minimum impact on the environment. Therefore, developing tree breeding Programmes to obtain quick gains and developing cloning systems to have a well established vegetative propagation method becomes important. The vegetative propagation methods should rapidly transform the genetic gains, obtained through breeding, into benefits for the industry. One of the most efficient tools to acquire these objectives is the combination of inter-specific hybridization and establishment of clonal forestry derived from superior hybrid individuals (Assis 2000).

In this context hybridization is an alternative of great impact in tree breeding Programmes, which can combine superior wood characteristics with tolerance to biotic and abiotic stress, thus representing a significant source of superior individuals, capable of introducing genetic gains in forest productivity and wood properties. Crossing species of different characteristics allows production of complementary wood properties in trees especially to meet industrial requirements.

Considering that Eucalyptus hybrids are heterogeneous, the effective and quick integration of genetic gains obtained with hybrids into the industrial process depends basically on existence of functional large scale cloning systems. Mass vegetative propagation perfectly complements hybridization for producing clonal forestry, and has some advantages over the sexual methods of mass reproduction of selected families, besides being the best way to commercially exploit the heterosis found in several Eucalyptus hybrid crosses. By capturing the total genetic variance (Zobel 1992), vegetative propagation allows for maximum benefits of wood properties and productivity, besides allowing for production of more uniform raw material, which from industrial point of view is highly beneficial to industrial process and product quality. Therefore, the tree breeding Programmes that focus on these aspects of forest industry will have great impact on the three important components of the competitive process: productivity, product quality and production costs. This article basically describes the recent evolution of cloning Eucalyptus for industrial purposes in Brazil, with emphasis on the cloning techniques and on the systems used for mass production of plant propagules.

NEW CLONING TECHNIQUES

Considering the consolidation of cloning as a tool for the establishment of productive Eucalyptus clonal forestry and its positive effects on industrial process and product quality, the evolution of the systems used for cloning Eucalyptus certainly could be predicted. The precursor of this development started in the early 80's when cloning Eucalyptus by rooting stem-cuttings reached industrial dimension (Campinhos and Ikemori, 1983). Since that time, the major constraint for general adoption was its applicability only to a small number of species and clones. Rooting stem-cutting was not suitable for a large number of economically important species, including those important for energy and charcoal, like E. citriodora, E. maculata, E. paniculata and E. cloeziana, and an important number of clones of rootable species had problems for commercial propagation. Most of the problems were associated with accelerated maturation process causing rapid loss of rooting-predisposition, and manifestation of topophytic effect. The phenomena of topophysis affects clones in different intensities, and is the main cause of intra-clone differences in growth and reduction of rooting ability. Franclet et al. (1987) emphatically pointed out that topophysis induced physiological differences and these differences can result in sufficient intra-clonal variability that can suppress the potential advantages of cloning. Another limitation of stem cutting was associated with alterations of root system architecture, leading to root deformation. In many clones such deformations prevented their full genetic expression, consequently reducing the ratio between selected trees and number of clones effectively used. Because such limitations of rooting stem-cutting, alternative methods were developed for commercial cloning of Eucalyptus species.

Micro-Cutting

Based on the work of Assis et al. (1992), this technique was developed in Brazil in the early 90's. The idea came up with the observations that rooting ability of stem-cuttings decreases with ontogenetic aging and the decline may be faster than reported in the literature. In E. grandis for example, the rooting competence decreased from the fourteenth node up (Patton and Willing, 1974), while it took longer in the E. deglupta. Assis et al. (1992) observed that clones of E. saligna, E. grandis and E. urophylla that had equally high proportion of stem-cutting rooting in vitro, showed differential levels of decline in the rooting percentage when managed in clonal hedges. This indicated that some factor related to clone growth, encompassing period between planting and cutting harvest (6 months), could be responsible for these differences. Preliminary tests done at Klabin Riocell (unpublished) showed, independent the species, almost 100% rooting of the very young mini-cuttings obtained from the cotyledonary leaf pair and the same tendency was maintained in the difficult to root species like E. citriodora, E. cloeziana, E. paniculata, E. dunnii, and E. globulus. However, with age, ranging from few days to some months, the cuttings harvested from such young plants showed a marked reduction in their rooting ability and in some cases such ability was totally lost. These observations suggested that the rooting potential reaches the maximum value at high juvenility level (mini-cuttings from cotyledons) and is similar in all species tested. But the decrease in the rooting ability with seedling age differed among species, which was similar to that found in the older materials in the field. This suggests that, at some stage, part of the juvenility obtained through rejuvenation process in vitro (Gonçalves et al. 1986) and/or on basal sprouts of cut adult tress (Hartney, 1980) is being gradually eroded during the growth of the clones in the clonal hedges.

Thus, hypothetically, the rooting ability of Eucalyptus clones ex vitro should increase if the "physiological distance" between the maximum juvenility stage, obtained in vitro, and the propagule collecting stage is reduced. To test this hypothesis Assis et al (1992) used shoot apex (micro-cuttings) of very juvenile micro-propagated plants of E. saligna clone as propagules. These micro-cuttings had 30% higher rooting than the stem-cuttings. In order to verify the results in a more representative sample, the rooting indices obtained by the traditional and the new micro-cutting method, in seven clones of E. saligna and five clones of E. grandis were compared. Considering all the clones, the rooting of micro-cuttings, on an average, was 18% higher than the rooting of the stem-cuttings, which amounted to an increase in the range of 6.3 to 44.6 percent points over the average. In general, higher increases in rooting percentage were observed in clones with lower rooting ability with the stem-cuttings.

Other results were obtained from several other trials established to define substrate, growth substances, environmental conditions for rooting etc. One of the most significant findings of this new technology was complete elimination of the use of growth substances usually required for the rooting of stem-cutting (Assis et al. 1992). These substances did not increase rooting of micro-cuttings, instead in some cases reduced it, indicating that the endogenous auxin concentration in the juvenile tissues was sufficient to promote rooting. Based on these results, a super-intensive system of Eucalyptus propagation ex vitro was established.The main feature of the technique is the use of juvenile plants or plants rejuvenated in vitro, as source of vegetative propagules. Shoot apices are used as micro-cuttings, which are placed to root in a glasshouse equipped with temperature and humidity control. The actual size of micro-cuttings is about 7 to 8 cm with two to three leaf-pairs. Presence of the shoot apex is important for quality of the root system, because its presence induces taproot-like system. The micro-stumps left after micro-cutting harvest, sprout rapidly producing new micro-propagules, which can be harvested for use within a period of 15 days in the summer and 30 days in the winter. Since its first use, the micro-cutting technique is improving continuously by incorporating new research findings. The evidence of evolution of this technique are well documented in publications of Assis et al. (1992); Xavier and Comerio (1996); Iannelli et al. (1996); Assis (1997); Wendling and Xavier (1999); Higashi (2000), and Campinhos (2000). The participation of dedicated scientists such as Professor Acelino Couto Alfenas (U.F.V –Viçosa-MG) and Professor Antonio Natal Gonçalves (E.S.A.L.Q.- Piracicaba-SP) among others, was of fundamental importance for the establishment of an efficient super-intensive cloning system for Eucalyptus species. The technical contributions, reciprocally exchanged through a pre-competitive development model and intense information exchange, were the bases for the fast evolution of this new concept of cloning Eucalyptus in large scale.

Mini-Cutting

Probably, the first initiative to root mini-cuttings in Eucalyptus was taken in the early 80's, by using shoots obtained from the thinning operation of rooted stem-cuttings (unpublished). The early trials were discouraging due to inconsistent results that varied from highly positive to negative. Later studies showed that such inconsistencies were caused by nutritional deficiencies of the mother plants at the time of mini-cutting collection. A similar system called "cascade propagation", used in France, is described by Chaperon (1987). However, in the 90's, after the consolidation of the micro-cutting as a functional propagation system, that mini-cutting system became commercially viable for Eucalyptus cloning. Higashi et al., (2000) developed a functional and efficient mini-cutting cloning system and several other researchers added significant contributions to the development of this technique (Wendling et al.2000; Xavier and Wendling, 1988).

Mini- and micro-cutting techniques are very similar in concept and operational procedures, differing mainly in the origin of the initial propagules. The micro-cuttings are obtained from shoot apices originating from micro-propagated plants, and the mini-cuttings come from axillary sprouts of plants cloned by stem-cuttings. After rooting of the first shoot, the two techniques are identical. In some clones using mini-cutting, some propagation cycles (serial propagation) are required to obtain the reactivation for acquiring full potential of rooting capacity. In micro-cuttings such propagation cycles comes naturally by monthly in vitro sub-culturing of explants. Micro-propagation is unnecessary for easy to root species because high levels of juvenility can be obtained easily by inducing basal shoots therefore in such cases mini-cutting is technically and economically feasible.

Despite the success of these techniques is believed to be related to maintenance of very juvenile stage (Assis et al. 1992, Xavier and Comerio 1996), new findings suggest that their high rooting potential are also related to the better nutritional status of the mini-cuttings. In general, mini-cuttings or micro-cuttings root better than juvenile stem-cuttings of the same clones produced in clonal hedges in field. Since both materials are juvenile, the mini-and micro-cuttings cultivated in well-balanced nutritive solutions root better as a result of better nutritional status. The rooting superiority of mini- or micro-cuttings may also be related to differential level of lignification in the two groups of propagules. Compared to stem cuttings, these micro- or mini-cuttings can be considered as "herbaceous", and using these techniques many complication associated with lignin formation and its concentration increase in tissues can be avoided.

Although, the two techniques are very similar, but received different names. The two terms mini-cutting and micro-cutting were designated, by convention, to systems of vegetative propagation that have their origin in rooted cuttings and micro-propagated plants, respectively. Thus a system originating from micro-propagated plants is termed as micro-cutting and that based on sprouts from rooted cuttings is called mini-cuttings. Nevertheless, there is a tendency to change this terminology to avoid confusion with the traditional term micro-cutting reserved for rooting of shoots produced in vitro. Therefore a system of vegetative propagation ex vitro based on mini-propagules could be termed as mini-cutting, independent of its origin, including those originating from a micro-propagation system. Therefore the term mini-cutting will be used to designate both techniques on this paper.

Advantages of Mini-Cutting

Compared to the traditional stem-cutting rooting, mini-cutting has many advantages leading to operational, technical, economical, environmental and quality benefits. Operationally, the Labour demanded and cost is markedly reduced, due to elimination of Labour intensive treatment with growth substances and many other operations required in an extensive outdoor clonal hedge management system. Many field operations like soil preparation, fertilization, irrigation, cultivation, weeding, pest and disease control, sprout transport etc are replaced by intensive activities in smaller indoor areas at much lower costs, where the amount of chemicals used is also drastically reduced.

The rooting ability of mini-cutting is much higher than the stem-cuttings. Although the benefits vary among the species and clones, in general in poor rootters species or clones with stem-cutting better compare with mini-cuttings, which may increase rooting by 40%. The main reasons for such increase are related to higher levels of juvenility and optimal nutritional content of the tissues, which improves the rooting predisposition and speed of root initiation. The rooting speed of mini-cuttings has two other important consequences in a commercial cloning Programme: the time of the plants indoor is usually reduced to half compared to rooting stem-cutting, thus considerably improve the use of the facilities and reduce the exposures of basal tissue of mini-cuttings to pathogenic fungi. At the initial and more susceptible stage, the faster reaction of mini-cuttings inducing rapid protector callus formation at the basal end provides protection from pathogen resulting in reduced fungicide application.

Mini-cuttings produce better quality root system with a tendency for a taproot-like system in contrast to the predominant lateral root growth habit in the stem-cutting system. Apparently the connection between root and stem tissues in the mini-cuttings is more suitable due to lower lignification of the tissues involved. This kind of root system appears to be more responsive to fertilization and is more resistant to environmental factors and considerably reduces intra-clone variation. As pointed out earlier, the topophysis effects observed in several systems of vegetative propagation are highly harmful to clonal uniformity. In this respect mini-cutting offers the advantage of physiological homogenization of the propagules producing a much more uniform forest.

EVOLUTION OF MASS PRODUCTION OF VEGETATIVE PROPAGULES

When the rooting of stem-cutting was first introduced commercially in Brazil (Campinhos and Ikemori (1983), the vegetative propagules for establishing clonal forests were obtained from field plantations. This management strategy required annually reserving large areas, with the only objective of obtaining cuttings for commercial rooting Programmes. Although the commercial plantations could be used to produce propagules, the high time-demand to gather wood and the vegetative propagules made this alternative unfeasible.

The clonal hedge based concept of intensive management model was introduced sometime later. This system demanded smaller area due to high shoot productivity per unit area (Carvalho, et al.1991; Higashi 2000). Despite being a tremendous advance for mass production of shoots, clonal hedges were still complicated to manage and poorly controlled. To meet the demand of clonal forestry, the system still required large areas, intensive Labour, and large amount of nutrients and water. The propagators were mostly dependent on the weather since the system was completely at the mercy of environmental conditions.

Mini Clonal Hedges

At the beginning of the last decade, the development of micro-cutting technology for Eucalyptus (Assis et al. 1992) materialized the concept of super-intensive management of producing vegetative propagules at commercial scale. Like for cloning methods, the advent of the micro-cutting substantially contributed to the progress made in systems for large scale production of vegetative propagule ex vitro. Originally the system was based on mini-hedges established through rooted mini-cuttings, grown in small containers (dibble tubes). This system provided a series of technical and economical benefits as well as good root quality (Assis et al. 1992; Assis, 1997; Xavier and Comério, 1986). Despite great advance over the clonal hedges in field, mini-hedges faced some limitations. The outdoor mini-hedges were still hostages of climate, and problems related to adequate maintenance of nutritional status and leaf diseases continued, especially during winter. The main problems were, reduced photosynthesis rate, reduced nutrient uptake and high levels of nutrient loss by leaching during period of excessive rainfall or even during irrigation. These limitation led to the development of indoor mini hedge system.

Indoor Hydroponics Clonal Hedges

The development of new technologies contributed to increase the efficiency of the existing systems. Hydroponics concepts for production of mini-cuttings was introduced first in operational indoor system based on drip fertigated sand beds (Higashi et al. 2000). The major contribution of this system was in terms of assuring an adequate nutritional status of the mini-cuttings, which is the key factor to obtain high rooting percentage in juvenile vegetative material. Using the same concept, Campinhos et al. (2000) developed a highly efficient method based on intermittent flooding system, where the containers of the mini-stumps are immersed in a nutritive solution for fertigation. These two hydroponics methods, sand-bed and intermittent flooding, are most widely used and the Brazilian companies use either of them.

Virtual Clonal Hedge

The observation that the shoots of first collection root better than those after third or fourth collection, suggested that vegetative propagules could be produced anywhere, without allocation of a definite physical area. Assis (1997) suggested a system where the mini-cuttings can be produced in virtual clonal hedges. Any rooted plant with normal growth when reaches a height of about 15cm, has its apex (7 to 8 cm) cut back to produce a new mini-cutting. As soon as the newly rooted mini-cuttings are about 15 cm high, their apices can be used as a new source of mini-cuttings, thus creating a virtual mini-hedge after some cycles of re-rooting, and requiring no specific area for production of vegetative propagules. The important attribute of this system is that the plants used for producing mini-cutting can be planted in field without losing quality of the intermittent flooding mini hedges.The sprouting of harvested plants becomes an extra option of providing mini-cuttings. The blanks resulting from rooting failure can be filled with mini-cuttings produced on the mini-stumps of the harvested plants. Similarly, the sprouts produced on the pruned plants and the apices of the rooted mini-cuttings can be used continuously as self-sustained cyclic process.

This system represents an excellent technical, economical, and operational alternative, but only works well if good nutritional conditions of the mini-cuttings are ensured. An extension of sub-fertigation to newly rooted mini-cuttings is the most suitable option to make virtual mini-hedging practical. The use of very soft mini-cuttings must be avoided to prevent fungal infection. The locations with wide climatic variations and without any environmental control are less suitable for establishing virtual mini-hedge system, therefore is not recommended in situations where environmental control is not feasible.

Advantages of Indoor Hydroponic Mini Hedging

The development of the concept of super-intensive hydroponic mini hedging, brought several benefits to commercial cloning of Eucalyptus. All indoor mini hedging systems have technical and economical advantages relative to normal outdoor clonal hedging. Higher productivity of mini-cuttings, lower Labour demand, and low consumption of chemicals and water, represents major sources of economic gain. In addition, considering the important effect of the conditions under which the mother-plants grow, on the rooting success (Read, 1987; Hartman and Kester, 1983; Haissig, 1986; Andersen, 1986), the main technical advantages are related to such aspect. These systems allow the use of important theoretical concepts which are difficult to implement outdoor, such as CO2 enrichment, control of temperature, light intensity and photoperiod. These factors in addition to the nutrition are of fundamental importance to enhance rooting predisposition. This system also enables application of pre-harvest treatments such as spraying mother plants with growth substance (phyto hormone) to increase rooting potential.

Propagules Productivity

Eucalyptus cloning in large scale started in Brazil using very extensive systems and the productivity of vegetative propagules was 114 cuttings/m2 of hedges field(Campinhos and Ikemori, 1983). With the introduction of the mini-hedges the production increased progressively to 121 cuttings/m2 (Carvalho et al. 1991), then to 1,752 cuttings/m2 per year (Higashi et al. 2000) and presently is at about 24,000 cuttings/m2/year. Table 1 provides an overview of the evolution of vegetative propagules productivity in Eucalyptus. The super-intensive systems, especially hydroponics based, are much more efficient, having productivity 350 folds higher than the initial systems used in Brazil. Comparing the different systems, the capacity of producing vegetative propagules in the super-intensive systems is highly expressive, consequently, more cuttings can be produced per year per unit constructed area of mini hedge, besides they are ease of handle, manage and they have a lower cost per plant.

Variations in the productivity of mini-cutting observed in table 1 resulted from different hedging systems, from different genetic materials and also from different management practices. Clones of E. grandis x E. urophylla hybrids are generally much more productive in mini-cuttings than other species. The majority of the pure E. saligna clones are much more difficult to produce as mini-cuttings than the E. grandis x E. urophylla hybrid where production/mini-stump is almost double. Still, the rooting competence and propagule productivity is much higher than in the outdoor clonal hedges. The other source of variation in the productivity is related to the concept and kind of management adopted to utilize the sprouts.

Certain practices of management of mini-stumps in the sand beds produce vegetative propagules similar to the macro-cutting system. The sprouts are allowed to grow to bigger sizes, and after collection each sprout is divided into three mini-cuttings (basal, middle and apical). This system produces higher number of propagules/m2 but it loses the best characteristic of the mini-cutting provided by the presence of shoot apex, responsible for producing high quality root system (taproot-like system, discussed earlier)

In both hydroponics systems (drip fertigation and intermittent flooding) losing mother plants is common due to salination, excessive harvesting, diseases etc. In this respect, the systems that use containerized mother plants, as in intermittent flooding, have advantage over the sand bed. In sand beds, the dead plant replacement and getting them to productive stage require more time compared to the intermittent flooding system.

The most attractive characteristic of virtual mini-hedges, that do not require allocated physical area for producing vegetative propagules, is that the mother plants are plantable. After supplying mini-cuttings, such plants are managed as normal rooted mini-cuttings. In the intermittent flooding system, the mother plants also can be planted provided the harvesting regime has not degraded its quality. Under successive and heavy harvesting regime the quality of mother plants frequently declines.One question that arises from the spacing in mini-hedges is about the influence of close spacing on the rooting of mini-cuttings and the mini-stump productivity. Trials at Klabin Riocell, showed that the spacing of 5 x 5 cm between plants does not interfere significantly in the rooting capacity of the mini-cuttings, but wider spacing permits higher survival rate of mini-stumps. But the final results are better with narrower spacing, since there is more than proportional compensation by existence of higher number of plants/m2.

Table. Propagules productivity in different clonal hedging systems used in Brazil*.

System	Species/ Spacing (m)	Productivity (Propagules/m²/year)	References
Clonal bank	E. grandis x E. urophylla - 3,0 x 3,0	114	Campinhos & Ikemori (1983)
Clonal Hedging	E. grandis x E. urophylla - 1,0 x 1,5	121	Carvalho et al. (1991)
Clonal Hedging	E. grandis x E. urophylla - 0,5 x 0,5	1752	Higashi et al. (2000)
Mini hedging Out door	E. grandis x E. urophylla - 0,05 x 0,05	29200	Xavier & Comério (1996)
Hydro mini hedging (sand bed/drop irrigation)	E. grandis x E. urophylla - 0,1 x 0,1	41480**	Higashi et al. (2000)
Hydro mini hedging (interm. flooding)	E. grandis x E. urophylla - 0,05 x 0,05	25500	Campinhos et al. (2000)
Hydro mini hedging (interm. flooding)	E. grandis x E. urophylla - 0,05 x 0,05	24000	Klabin Riocell operational
Hydro mini hedging (intermittent. flooding)	E. saligna and E. urophylla x E. globulus - 0,05 x 0,05	14400	Klabin Riocell operational
Virtual mini hedging	E. saligna and E. urophylla x E. globulus 0,05 x 0,05	-	Klabin Riocell operational

*Based on cuttings/mini-cuttings/stump considering 12 months of effective production, without discounting the replacement of dead plants.

**Subdividing sprouts in 3 mini-cuttings

Temperature and Light

Temperature and light are good examples of the factors that can be controlled in mini hedges, to improve rooting predisposition of mini-cuttings. Light can be managed as to photoperiod and intensity. Light intensity can strongly influence cutting productivity and rooting by reducing or increasing endogenous phenolic substances that act as rooting inhibitors (Vieitez and Ballester, 1988) or promoter depending upon the concentration in the tissues and the species involved. The photoperiod, especially combined with temperature can influence rooting predisposition of shoots. Lighting also can influence rooting by controlling internal levels of carbohydrates. Etiolation of mother plants can affect rooting and root number, which depending upon the specie may increase or decrease (Hansen, 1987). This variation appears to be linked to carbohydrate production (Nanda et al. 1971) and accumulation and transport of auxins (Hansen, 1987).

Temperature can influence rooting by interfering with nutrient uptake and metabolism, and its control, especially in sub-tropical climates, can be adjusted for optimum production of cuttings. Assis (1997) observed that rooting percentage of mini-cuttings decreases in cold winter months. Recent work at Klabin Riocell (Unpublished) showed that this problem could be solved by supplying additional light (14h/1000 lux) and increasing the temperature in environment of mother plants (>20 oC). Since the two factors were not separated, the effect of individual factor could not be determined. However, considering that nutrient concentration is of prime importance for rooting process, and that their uptake depends on the metabolic activity, it can be assumed that both the factors contributed to reestablishment of normal rooting competence. In general, in areas of cold climate, rooting is limited to only warmer months, therefore as discussed above, if light and temperature are controlled the use of greenhouse space can be optimized by extending the rooting period for the entire year.

Nutrition

The most important advantage of hydroponics mini-hedges is the possibility of providing well-balanced nutrition to mini-cuttings. The effect of macro- and micro-nutrients on root initiation of cuttings on several plant species, has been well documented in the literature (Andersen, 1986; Read, 1980; Hansen, 1987; Haissig, 1986; Hartman and Kester, 1983). As discussed earlier, nutrition is the key factor affecting rooting predisposition, because of its involvement in determining the morphogenetic response of the plants. In hydroponics mini-hedging system the nutrients are supplied in ideal concentrations to promote rooting. Hydroponics, without the influence of rainfall and or overhead irrigation, allows better nutrition control of mother plants. The use of indoor hydroponics mini-hedges allows nutrient supply in a correct concentration, avoiding the nutritional imbalance commonly observed in outdoors clonal hedges; besides there is no leaching by rainfall and the response to nutrient application is rapid. According to Haissig (1996), a feasible nutritional balance of mother plants can be linked to production of tryptophane, the precursor of IAA and to storage substances.Calcium plays an important role in the rooting process, mainly because it acts as peroxidase activator (Haissig, 1986). According to practical experience of Klabin Riocell, the percentage rooting of mini-cuttings decreased considerably if the leaf calcium concentration dropped below 0.7%. This situation was observed especially in winter when, in the absence of heating, leaf calcium concentration was substantially lower than in warmer months. Calcium deficiency induces shoot apex necrosis in mini-cuttings, which can cause problems of rooting and mini-cutting development (McComb and Sellmer, 1987). This phenomena was commonly observed on the initial, operational scale, outdoor mini hedges.

In general, moderate deficiency of nitrogen can improve rooting (Haissig, 1986), but at high levels more energy is required for vegetative growth and especially for leaf expansion. Consequently, carbohydrates are not stored at suitable levels reducing the C:N. On the other hand, extreme nitrogen deficiency can reduce rooting since it is necessary for nucleic acid and protein synthesis. Based on the considerations of Hartman and Kester (1983), and Haissig (1973b, 1986) a general model for leaf nutrient concentration can be established. The mother plant must be well nourished with macro nutrients like phosphorus, potassium, calcium and magnesium and moderately deficient in nitrogen.

The effects of micro-nutrients on rooting are variable. Manganese, due to its inhibitory effect should be used at minimal concentrations, but boron and zinc are essential for rooting process (Andersen, 1986). Zinc increases the endogenous content of auxins by increasing trpytophane proportions while boron is involved on rooting and root growth. Table shows nutrient concentrations in leaves of well nourished plants (Roberto Ferreira Novais, pers. Comm.) for rooting mini-cuttings. Monitoring leaf nutritional concentration of the mother plants, as a guide to assure these levels of macro and micro-nutrients has been an important factor in improving the rooting competence of mini-cuttings.

Table. Concentrations of macro- and micro-nutrients in leaves of well nourished Eucalyptus plants.

MACRO NUTRIENTS (%)					MICRO NUTRIENTS (mg/kg)					
N	P	K	Ca	Mg	B	Mn	Zn	Fe	cu	S
2,5 to 3.0	0,2 to 0,4%	1,5 to 2,0	1,0 to 1,5	0,25 to 0,40	40 to 70	100 to 500	50 to 60	100 to 200	10 to 15	0,15 to 0,25

Phytosanitary Advantages

Shoot contamination by splashing of soil during rainfall or over-head irrigation (Ferreira, 1996) is a very well known sanitary problem associated with outdoor clonal hedges. As many pathogenic fungi are soilborne, the cuttings produced either in clonal banks or in outdoor clonal hedges, are frequently contaminated by these fungi. Thus one of the advantage of using indoor hydroponics is the preventive phytosanitary control. In the absence of leaf wetness occurrence of leaf diseases is greatly reduced, thus producing pathogen free cuttings that increases the results of rooting. However, all these benefits of indoor system can be lost or become disadvantageous in the absence of efficient controls to minimize effects of external factors such as extreme temperature and humidity.

MANAGEMENT OF MINI-STUMPS IN HYDROPONICS SYSTEMS

The management of mini-stumps is rather simple, but some aspects must be considered for the technique to be successful. Mini-cutting collection should be done selectively to avoid mini-stump degradation. Desalination should be done at least once a week by generously irrigating the substrate to leach the surface-accumulated salt. In non-automated nutritive solution preparations, the entire solution should be changed every 15 days. In intermittent flooding better results are obtained when the electric conductivity is adjusted to 1,8 – 2,2 in the winter, and to 0,8 – 1,0 in the summer. Avoiding cut and abscised leaves to fall into the nutritive solution (Alfenas, 2001) prevents biotic and abiotic problems on mother plants.

USE OF MINI-CUTTING FOR CLONING OTHER WOODY SPECIES

The applicability of mini-cutting technique has been tested also on other, broad leafed and conifers woody species. In general, the methods used are similar to those of Eucalyptus, with minor adaptations. The results have shown that this technique can be fully applied to species like Pinus taeda, P. elliiottii, Acacia mearnsii and Ilex paraguaiensis and many other woody and non-woody species. The best results were obtained with juvenile plants as is the case with difficult to root E. citriodora, E. maculata and E. paniculata. Thus, mini-cutting is feasible to be used consistently for family multiplication of these species. Use of mini-cuttings can be a possible alternative for production of family forests by multiplication of full-sib or tested half-sib families.

Klabin Riocell started a small Programme of establishing E. globulus clonal plantation, based on adult and early selected trees resistant to rust caused by Puccinia psidii and other diseases. The preliminary results suggest that mini-cutting is a very promising technique to commercially propagate this species. Compared to stem-cutting, the mini-cutting gave better results in rooting proportion and root system quality. It appears that the propagation of E. globulus is profitable, since it is considered to have different and higher nutritional requirements. The development of special nutritive solutions for E. globulus tends to improve the results and it is clearly possible to establish efficient clonal Programme for this species.

In Acacia mearnsii, mini-cutting has shown good results, despite being considered as recalcitrant species with regard to vegetative propagation (Matheson 1990). Despite reasonable success obtained by Assis et al. (1993) in adult trees, application of mini-cuttings gave better results, indicating possibility of developing clonal forestry of this genus. The key point in this case is to achieve juvenile or rejuvenated shoots to start the process. Through induction of basal shoots, without tannin oxidation normally present in the chopped trees of this specie, it is necessary to establish the initial healthy rejuvenated plants, which should done using standing trees to avoid problems with tannin.In Pinus, only juvenile seedlings were tested as mother plants. Pinus elliottii seedlings can be easily rooted by mini-cuttings with more than 85% success. Pinus taeda also roots well though somewhat less (76%) than P. elliottii. The results suggest the possibility of developing forestry families Programmes with both species based on the systems used for Eucalyptus. Managing Pinus mini-hedges in hydroponics is efficient for providing a good nutritional status to the mini-cuttings. The productivity of mini-propagules, without any change in the nutritive solution, was 9600/m2/year. The results of mini-cutting productivity and rooting proportion are highly encouraging.

9

Genetically Engineered Trees in Forestry

INTRODUCTION

Intent

This paper will address the question of whether there will ever be the certification of forests that contain genetically modified (GM) trees. This chapter, chapter two, reviews the origin and range of both certification systems and genetic modification (GM) applications, and defines a genetically modified organism (GMO). The third chapter will focuses on the aspects of GM that certification schemes, and the Forest Stewardship Council (FSC) in particular, find problematical. It doing so it will consider how the different certification schemes have addressed comparable issues to those raised surrounding GM trees. This includes the conservation of genetic resources, use of exotic species and application of chemicals. Chapter four briefly reviews the inherently different perceptions of GM by the various stakeholders, and the role this plays in risk evaluation. Lastly, chapter five concludes with consideration of the politics and business surrounding GM and certification, and reflects upon their greater role in determining whether forest certification and genetically engineered trees will ever be compatible.

The rise of certification

International responses to deforestation have been vociferously criticised by many NGOs . These reproaches centre on the perception that there has been a slow implementation of inadequate forest conservation agreements. Furthermore, these agreements themselves do not commit participants to agreement implementation. These deficiencies have been recognised for some time; an FAO report on TFAP noted an "inadequate regard for local peoples, ecological matters and the underlying causes of deforestation" and recommended "avoid bureaucratic suffocation and encourage effective leadership" (Ullsten et al. 1990). The ITTO has been consistently criticised as unrepresentative with ineffective environmental concern and poor management, ignoring their prescribed objectives, such as Target 2000 (Colchester 1990; FoE 1992). Moreover, the impasse between 'North and South' has, in the eyes of many NGOs, lead to painfully slow progress since the UNCED 1992 summit; many countries have yet to ratify agreements, and international initiatives (IPF, IFF, ITFF and UNFF) established to further UNCED work, have made slow progress (Humphries 2000).

However, from these processes, SFM emerged as a new framework. It aims to describe management that ensures long-term forest health and productivity, while providing continued social and economic benefits (Evans 1996). In 1988, the ITTO noted that only 0.08% of tropical forests were managed in this manner (Poore 1989). Although endorsed in principle, SFM is not well understood in operational terms. Currently, principles, criteria, and indicators are being developed as a means to assess and report on progress towards SFM. These have been developed largely through international agreements building on UNCED 1992; principally the Helsinki Process, the Montreal Process, the Tarapoto Proposal, a renegotiated ITTA and Dry-zone Africa (Grayson and Maynard 1997).

It has been noted that deforestation continues apace, and legislative measures to curb unsustainable measures have largely failed due to international squabbles, general apathy and agreements that are non-binding (Dunleavy 1993). Further, the de jure appearances of legislation are seen to ignore the de facto realities of implementation at the forest level. A market-based mechanism of certification was considered as a means of circumventing these failings. Being site-specific, certification could validate 'on-the-ground' operations as employing the best management practices through criteria and indicators (Upton and Bass 1995). Consumers, using a labelling system attesting to this SFM, were to discriminate between different production methods. Those companies who were certified could gain many benefits, chiefly improvement / retention of market share, defence against environmental criticism, and investor assurance. This would drive the implementation of SFM in the wood product trade. Those who continued to practice and trade in uncertified products would have difficulty marketing their produce. Credibility was seen as a precondition for any successful certification scheme; independence and third party assessment were understood to be requisite (Upton and Bass 1995).

Certification is a voluntary process, which results in a written statement attesting to the origin of wood raw material, and its status following validation by an independent third party (Ghazali and Simula 1994). Certification typically includes two main components: forest management certification and product certification. Forest management certification is based on an assessment of forest management against a set of standards reflecting contemporary concepts of sustainability. Product certification involves verifying the chain of custody of wood from the certified source to the consumer. This whole process has faced repeated criticism (Counsell 1996; Centero 1998). Nevertheless, governments and companies are examining certification with interest; it has become a "high profile subject in the forestry sector" (FAO 2000a). Despite the stated purpose of certification for improving forest management, the main interest of most of those undertaking certification at present is probably the marketing benefits it may offer. This may explain why over 80% of FSC-certified forests are in developed countries, and 66% are by industrial enterprises (Thornber 1999). However, in developing countries it is noted "certification serves as an added strength as it facilitates entry into foreign markets" (MTB 1999).

Increasing numbers of certification schemes are being developed; international, national and regional, all can be split into two types.

- The performance based approach, best represented by the FSC and some emerging national schemes (Indonesia, Malaysia and Finland), is founded on standards that an organisation has to meet before it can be certified. This method has been criticised because, although certification is an incentive for forests that are already well managed, companies managing forests poorly may find adjusting their practices very difficult. Consequently they may simply ignore certification (von Maltitz 2000).
- The process-based approach, which is the basis of the system adopted by ISO 14000 series of EMS verification, assesses the quality of an organisation's management process; how it sets its policy and management objectives, and how it organises itself to deliver them consistently. They do not lay down specific pre-determined performance standards, but rather look for continuous improvement of environmental management. The ISO 14000 series, and in particular the forestry-specific 14061, ensure companies set increasingly higher standards and achieve them. This approach is criticised as a means of gaining certification by compliance with internal policy, whilst still being environmentally irresponsible. Moreover, 'ring fencing' can give a misleading impression that ISO standards are applicable to the whole of a company. Lastly, since ISO places a great emphasis on achieving national standards, it could be argued it maintains the disparity in performances between countries, and consequently misleads consumers (von Maltitz 2000).

Technically, ISO certify management systems and not forests, and as such they do not enable product 'labelling'. It is probably for this reason that ISO implementation is greater in companies supplying predominantly to the pulp and paper sector, whilst companies supplying wood timber opt for performance-based systems, such as the FSC. For example, all SAPPI's plantations in South Africa are ISO-certified, but only those supplying saw-logs are FSC-certified (von Maltitz 2000). This contrasts with nearby Mondi plantations, with over 430,000 ha of FSC-certified timber forest producing only saw-logs (FSC 2000e). ISO certification meets pulp and paper consumer demands, and thus serves as an adequate marketing tool. However, ISO has been used by a number of companies as a step towards gaining product certification (Bass and Simula 1999).

Increasingly, process-based schemes are incorporating performance targets; correspondingly, performance-based schemes acknowledge the benefits of EMS (Kanowski et al. 2000). Many see the FSC and ISO approaches to certification working in tandem , complimenting international and national policies to ensure SFM is implemented at all levels. There is little support for the idea of harmonisation of certification systems, but growing support for mutual recognition. For example, six certification systems, including SmartWood (US Rainforest Alliance) and Woodmark (UK Soil Association) are harmonised with the FSC, and their standards are mutually recognised.

Prominent NGOs support the FSC process because they consider its standards to be the most stringent and open too little 'interpretation'. These supporters are concerned that mutual recognition should not diminish FSC standards. However, many advocates of alternative certification initiatives interpret this concern, rightly or wrongly, as an attempt to maintain the dominant role of the FSC in certification. Regionally-based systems are often espoused as better tailored to meet inherently different local circumstances, and commercial plantations perceive the FSC as bias towards natural forest management. However, the FSC has committed itself to collaboration with Malaysian and Indonesian certification initiatives, with the eventual objective that certification, by either party in those countries, might be recognised by both parties (FSC 1999d).

The development of genetic modification (GM)

Paralleling the rising concern over deforestation, much scientific and commercial attention has focused on improving the genetic stock of forest trees in order to improve productivity and quality. The demand for wood-based products is forecast to double over the next 17 years, the onus being on paper and board products (Soil Association 1998), which must be met from plantation-grown trees. GM and other biotechnology techniques have provided conventional tree breeders with new tools to meet this demand, enabling the accelerated modification of some forest trees possible (Dinus and Tuskan 1997). The potential gains are great, since trees remain genetically and phenotypically very similar to their wild progenitors (Wright 1976) and GM can progress more rapidly than conventional breeding, which can take decades.

The first commercially interesting genes available to GM in forestry were those which were developed in agriculture, and thus involved the transfer of

DNA between taxonomically distinct organisms. Traits currently being researched include herbicide resistance, increased vigour, pest and pathogen resistance, increased tolerance to biotic stress and improved timber quality (Riemenschneider et al. 1988; Bauer 1997; Dickson and Walker 1997; Tzfira et al. 1998).

Recently, particularly in Europe, GM has been at the centre of a heated debate. This has concentrated on agricultural applications, but has occasionally touched forestry. The issues are enormously complex and not related to purely scientific questions: there are questions of philosophy, ethics, equity, responsibility and ownership. Most proponents have called for "credible information that promotes rational debate" ensuring the public can make an "informed evaluation" (Strauss et al. 2000b). Some of the media have also put forward what appears an assessment of potential benefits and risk (Weiss 2000). However, the media are usually criticised for hyperbole. In the UK, "pseudoscience" and "alarmist media reports" from pro- and anti-GM camps have confused matters and brought GM into the limelight (Nuffield Council on Bioethics 1999). Opinions are sometimes extreme. Some call for deregulation and accelerated introduction of GMOs for economic and environmental reasons (Cantley 1998), whilst many environmental groups are very wary of potential impacts, and consequently believe "genetically modified organisms must not be released into the environment" (Greenpeace 2000).

The UK is an interesting case. The public generally perceive the attitudes of the large companies pursuing GM as irresponsible and principally profit driven. Furthermore, consumer confidence in food is very low following episodes of salmonella in chicken eggs, E. coli and BSE in beef, and the reprocessing of condemned poultry for human consumption. Coupled with extensive media coverage, these two factors have acted synergistically, and the public are sceptical of the motivation behind, and safety surrounding, GM. On 11th July 1999, campaigners destroyed most of the trees in a research trial of GM poplars established by AstraZeneca. This action illustrated how strongly some people felt about the potential ecological impacts of GM. These impacts include gene flow, invasiveness and unpredictability (Owusu 1999). However, there are also potential ecological and economic benefits (Mathews and Campbell 2000).

The advantages and disadvantages of GM trees have been considered by many (reviewed by Mullin and Bertrand 1998; Mathews and Campbell 2000), but as yet no GM trees are in commercial production. Big-business is regarded by many as an outright advocate of GM plants (Allen 1999). Conversely, the FSC and its allied certification schemes, advocating environmental campaigners concerns, have banned outright the use of all GM trees in the forests they certify (FSC 1997, 1999b, 2000a; Soil Association 1998). However, many scientists believe there is a middle-ground between these two views, and have noted "the opponents of the technology have framed the issue as black and white. GMOs are dangerous and must be stopped. Proponents are faced with the difficult task of trying to educate the public about the many shades of grey" (Somerville 2000). This has raised debate about the rationality behind a complete ban on GM trees, for "each GMO should be assessed on its own merits" (McHughen 2000, p.113).

GM THROUGH CERTIFICATIONS EYES

Certification bodies can be placed in two groups; against GMOs and tolerant of GMOs. The first camp is centred solely on the FSC; Principal 6.8 of FSC guidelines states the "use of genetically modified organisms shall be prohibited" (FSC 2000a). This is reiterated in national standards; line 266 of the FSC Forest Management Standard the United Kingdom states simply insists "genetically modified organisms (GMOs) are not used" (FSC 1999b), as does 6.5.11. of the Swedish FSC Standards (FSC 1997). Certification systems recognised as equivalent to the FSC, such as The Soil Associations' Woodmark (section 5.610) also prohibit GMOs.

For many large companies considering certification, the FSC is one of the international systems, largely due to its international scale and backing from influential NGOs such as WWF, Greenpeace and Friends of the Earth. Although the FSC have yet to publish an official document qualifying their policy towards GM , supporting NGOs have disclosed their concerns (Owusu 1999) and some resolutely reject the use of GM plants (Greenpeace 2000). This is important because these NGOs are considered to hold substantial influence over FSC policy . It is likely FSC policy will parallel much of their supporters' concerns, both now and in the future.

The Soil Association published their policies regarding GM, and note that "it would be sensible to apply the precautionary principle and not release GMOs into the environment, at least till their long term impact has been determined" (Soil Association 1998). When discussing certification accreditation they state "this position will be kept under constant review but the evidence that is continuing to emerge confirms the need for prudence and reinforces the logic of adopting a precautionary principle in conjunction with a comprehensive programme of research and monitoring" (Soil Association 1998). Thus, if research and monitoring of GMOs satisfy certifiers' concerns, GM and certification may be compatible, despite the current FSC stance regarding GM.

The other certification camp is more varied. Although it claims not to be certification system, the SFI standard of the AF and PA has all the trappings of, and is considered by others to be, a certification system (Kiekens 2000; McKeand, pers. comm.). This U.S. programme is industry-lead and supported by a number of NGOs, although not nearly with the profile of those that back the FSC. However, the SFI Programme encompasses more than 26.8 million hectares (Patrick 2000), and by the end of 2001 AF and PA projects that 12.1 million hectares of member company forestlands will have been third-party certified (Pulp and Paper 2000). This compares with just 1.8 million hectares of US forest currently certified by the FSC (FSC 2000b).

What is a GMO?

We lack a precise and common definition of a GMO. The FSC adopted a definition that combined principals outlined in EC Directive 90/220 with the UK Government Health and Safety Executive publication on Contained Use of GMOs (EU 1990, UKGovernment 1996). This indicated a "Genetically modified organism (GMO) means an organism in which the genetic material has been altered in a way that does not occur naturally by mating and/or natural recombination or both " (FSC 1999a). This definition encompasses DNA introduction via living vector systems, such as Agrobacterium tumefaciens used with poplars and eucalyptus (Griffin 1996), and mechanical DNA introduction methods, such as biolistics, used with Pinus radiata (Sederoff and Stomp 1993). Moreover, 'unnatural' cell fusion and hybridisation techniques would be classified as GM. All other technologies were considered permissible.

The continuum and complexity of different breeding technologies makes distinguishing between 'traditional' breeding from genetic modification very difficult. Humankind has been genetically altering crops for millennia (Diamond 1997). This has resulted in varieties almost unrecognisable from their wild progenitors, and these changes may be very rapid. For example, the apical dominance of maize (Zea mays spp. mays) makes today's cultivars strikingly different to their wild progenitor, teosinte (Z. mays spp. parviglumis). This change results largely from alteration of the regulator sequence of the teosinte branched1 (tb1) gene (Doebley et al. 1997). The remainder of the tb1 gene is identical in both maize and teosinte (Wang et al. 1999). Moreover, the distinction between breeding systems may be inconsistent with the FSC's stated concerns over GM (FSC 1999a).

For example, herbicide resistant plants have been derived both 'conventionally' and through GM (Concar 1999). The potential risks of gene flow and enhanced weediness from such cultivars are considered to be very similar to GM cultivars. Indeed, some accepted methods may be riskier than GM. For example, the FSC notes specifically that mutagenesis is permitted (FSC 1999a), a process that involves exposing cells to ultraviolet light, radiation or some of the "most poisonous chemicals known to humanity" (McHughen 2000, p.65). The genetic changes this induces are usually not understood. GM has knowledge of exactly which genes have been introduced, and how they function (MacKay et al. 1999). Leading research panels have concluded that the genes and traits must be examined, rather than the method of genetic modification (National Academy of Sciences 1987; National Research Council 2000).

This dichotomy between anti-GM and pro-GM camps reflects an underlying philosophical concept, 'natural' and 'unnatural' , and it is crucial to understanding the debate about biotechnology (Kershen 2000). The label applied to a product depends on the 'view' of the concept of 'natural'. One influence on this view may be the FSC themselves, yet other possible influences include the views of those who lobby the FSC.

It must be remembered that certification is a marketing exercise to help promote SFM, and the FSC must create a brand, a label, respected by the consumer. The consumer must believe that certification testifies to sustainable forest management. Thus, the view of 'natural' is imperative. If the FSC feels more akin to environmental groups or more politically influenced by these groups or a 'green' public, the FSC may endorse the 'nature' view with concomitant consequences for regulatory policy; GM will not be certified since it is an 'unnatural' artefact of man's activity.

It is this philosophical divide which is paramount in the debate surrounding genetic modification . GM is viewed by many 'environmentalists' and members of the public as intrinsically different from other advanced breeding techniques, including mutagenesis, embryo rescue, somaclonal variation and cell selection. This may reside in the ability to insert genes from taxonomically distinct species . It is not entirely rational, since many crosses between plants of the same species involve plants grown so geographically dispersed that they would never mingle DNA without human intervention.

However, the ecological impact of producing hybids and intoducing exotic species are cited by some as a case for the possible environmental impact of GMOs (Owusu 1999). The need for objective evidence is required by both sides of the debate, and substantive equivalence should be invoked when considering how valid the concerns surrounding GM are. Even the impact assessment of the risks of GM is fraught with philosophical arguments. As noted by Apel (2000), it is necessary to focus on those factors relevant to an objective risk assessment, based on scientific reasoning. However, it is not clear how this might be accomplished in practice.

THE SPECIFIC CONCERNS OF CERTIFICATION BODIES

The FSC are the only forest certification body to have banned the use of GM trees in the forests that they certify (FSC 2000a). This ban was imposed due to concerns about the environmental safety of GMOs. The concerns have been published in an unofficial document (FSC 1999a), and most are identical to concerns expressed by environmental NGOs (Soil Association 1998; Owusu 1999; Greenpeace 2000). The following chapter focuses on the FSCs stated concerns about GMOs, and considers each issue by discussing equivalent practices examined by the FSC. Where appropriate other certification systems are considered.

Reduced 'diversity'

GM technology is likely to be employed by fine-tuning genera and species already adapted to a site and which can be mass propagated for direct use in plantations (Griffin 1996). This has caused concern within the FSC: "Plantations using one or few transgenic clones will contain less landscape-level diversity than is currently found in plantations using species or varieties resulting from traditional tree-breeding" (FSC 1999a). Yet this misgiving is not restricted to GM per se. Clonal forestry is already well established for Eucalyptus and Pinus radiata, and has been performed with Populus species for over one hundred years (Muhs 1993). The risk of reduced diversity is equally applicable in any clonal plantation. Diversity requirements should be established regardless of whether a tree is GM, clonal or exotic; certification standards recognise this.

PEFC guidelines for Forest Management Practices aim to maintain and 'appropriately' enhance (PEFC 1998). What exactly is meant by genetic diversity is not made clear in this document. Furthermore, these guidelines expounded very little at a national level (PEFC 1999). For example, specific targets of genetic diversity are not set other than for natural regeneration. The SFI system is even more vague, and does not set out criteria for maintaining or enhancing genetic diversity (AFandPA 2000). In contrast, the FSC is specific: "Diversity in the composition of plantations is preferred, so as to enhance the economic, ecological and social stability. Such diversity may include the size and spatial distribution of management units within the landscape, number and genetic composition of species, age classes and structure" (10.3, FSC 2000a).

At a national level, the FSC sets detailed targets for the minimum species diversity in any plantation. For example, line 365 of FSC UK guidelines (UKWAS 2000a) limits the proportions of different species within a wood. A maximum of 65% of the wood may be of a primary species, and only 75% of the wood may be composed of one species if it is the sole species suited to the site and the owners' objectives. The remainder of the wood must be a certain combination of native broadleaves, open space, secondary species or areas actively managed for biodiversity. However, this has resulted in oak (Quercus robur and Q. petrea) coppices in Wales, S.W. England and the Wyre Midlands having to be diversified despite being the result of decades of traditional management (Simon Pryor, pers. comm.). Similarly, in Germany, forests that are considered models of excellent SFM cannot gain FSC certification because beech (Fagus sylvatica) dominates the canopy, just as it would under natural conditions (Hans-Albrect Wiehler, pers. comm.). Critics of such an arbitrary figure are justified; no account of the natural forest composition is made. Moreover, multiple species composition is not always correlated with diverse flora and fauna – woods with a high proportion of English oak (Q. robur) will have much greater biodiversity than a dense plantation of mixed exotic conifers (Peterken 1992). Each situation should be examined on its own merits.

In other words, all options should be explored when dealing with a problem, and in some instances chemical methods are appropriate because they can be economically or ecologically advantageous. Continuing this argument, GM should be considered, for it may produce cheaper and more environmentally sound pest control (USDA 1999).

The DSS gives guidelines about assessing which methodology, whether chemical or otherwise, to apply to potential problems. It outlines parameters such as cost, effectiveness, toxicity and impacts of different practices. It continues "If the use of a chemical is necessary and a choice is available, evaluate the risks of environmental damage". Further, "...reference should be made to 'Environmental Impacts' which gives general information and a table to assist in choosing the material with the least impact". Essentially, the DSS gives a permitted list of options, chemical or otherwise, and asks the forest manager to make a value judgement for best practice and damage limitation: "Users should decide ... using their professional judgement" and "determine the range of suitable pesticides, dose rates and application patterns from the tables and supplementary literature ... make an assessment on the likely effects of the pesticide using the table in this section". Thus, each individual situation needs consideration in the field, from which an appropriate course of action can be taken.

One could argue that in many instances GM fulfils DSS requirements, for the "aim should always be the minimum quantity of herbicide required to give the desired degree of control". An ERS/NASS Agricultural Resource Management study (USDA 1999) on herbicide treatments on GM herbicide resistant cotton and soybean compared application rates to all other seed technologies, and found that 50% to 60% of spayed crops had significantly lower herbicide application at the 5% level (USDA 1999). Further USDA research has shown that between 1997 and 1998, U.S. national pesticide treatments were reduced by 6.2% and was accounted for by the introduction of Bt resistant crops (Fernandez-Cornejo et al. 2000). Reduced herbicide applications may also lessen the selective pressures on weeds.

The DSS report continues "herbicides that have low toxicity should be favoured" and "in many cases a carefully directed spray of broad spectrum product will be the most effective option and offer the least risk to non-target species". These are often cited as pro-GM arguments. The introduction of a herbicide resistant transgene such as Roundup Ready® confers resistance to glyphosate, and has been considered as one of GMs greatest successes (Tzfira et al. 1998). The DSS states glyphosate (Roundup Pro Biactive®) is "not hazardous, not toxic to invertebrates, not harmful to aquatic life, and has low selectivity". It is a safe, broad-spectrum herbicide, considered environmentally benign because it breaks down quickly into non-toxic end products. If used properly, a Roundup Ready® gene could lessen the environmental impact of weed control; over the top spraying becomes feasible (Dickson and Walker 1997), enabling post-emergence treatment when weeds are vulnerable (Rogers and Parkes 1995), and a reduction in herbicide usage because the "use of pre-emptive weed control is often more effective than dealing with greater problems later" (UKWAS 2000b). A recent study confirmed these benefits; Roundup Ready® agricultural crops enabled the substitution of most synthetic herbicides with glyphosate. "The herbicides that glyphosate replaces are 3.4 to 16.8 times more toxic .... and persist in the environment nearly twice as long" (Fernandez-Cornejo et al. 2000).

Indeed, conventional weed treatments, such as tilling and mowing, can be very damaging to the environment. The DSS report notes that "Mowing is largely ineffective as a weed control measure" and "often needs to be repeated several times a year for many years". Furthermore, "mowing creates a grassy weed flora that is harmful to trees" and "can result in soil compaction ... and pollution from exhausts and spillage of fuel and lubricants". The reduced need for tillage control of weeds can lessen erosion (James et al. 1998).

The FSC recognises that, under some conditions, herbicide use is warranted. Chemical use is a judgement made by the forest manager. If he or she justifies usage soundly, having considered all possible alternatives, then certification follows. This rationale could be extended to GM. There are many possible applications of transgenics. The specific trait and its proposed use should be examined, just as for other technologies. Furthermore, substantive equivalence should be cited; the DSS goes on to state "there are some situations in which complete removal of all other vegetation over the whole site may be considered". Despite the fact that such a radical measure could have a significant effect on a site, it has been judged to be acceptable. It is highly unlikely that GM would result in removal of all other vegetation from a site, yet it has not been judged to be acceptable.

Insect resistant GMOs

The FSC consider a potential hazard of GM trees to include "increased resistance of target insect pests, and/or deleterious effects on natural enemies of the target insects, and/or deleterious effects on insects such as butterflies, pollinators, soil microbes, arising from the use of GMOs with insect resistance" (FSC 1999a). These concerns are thought to stem largely from U.S. studies on the effects of GM-corn modified with Bt transgenes. Two studies have reported negative impacts of Bt toxins on the monarch butterfly, Danaus plexippus; Losley (1999) and Hanson and Obrycki (2000). These papers have since been criticised as unscientific or taken out of context by the media (Shears and Sheldon 2000), but have generated public apprehensions. Other concerns are voiced by NGOs; Section 2.4 of a WWF report considers insect pesticide resistance, noting "Bt crops will augment the selection pressure placed on target pests and this will inevitably lead to an increased frequency on Bt-resistance genes within the insect's gene pool" (Owusu 1999). Such misgivings over continual plant insecticide production are valid. Bt sprays have been used for decades in conventional and organic agriculture, and are considered "a safe and effective bio-pesticide" (Greenpeace 2000), despite scientific research questioning the safety of Bt spays (Swadener 1994). For forestry operations in the US and Canada, Bt is often the only insecticide that is sanctioned. It has also been widely used in Europe; in Spain Bt is frequently applied against gypsy moth Lymantria dispar (Speight and Wainhouse 1989). However, the continued exposure of pests to Bt toxins has selected for many resistant insects (Tabashnik et al. 1990; Talekar and Shelton 1993; Tabashnik 1994; Bauer 1997; Tang et al. 1997; Speight et al. 1999). Such resistance is thought to have developed through continued exposure to sprays on non-transgenic crops, and where these sprays have persisted in the soil following application (Saxena et al. 1999).

Despite ten years of proven insect resistance in the field, Bt sprays remain on the market and are an acceptable practice under rigorous 'organic' guidelines (EU 1991). This may lie in the perception that Bt is a 'natural' product, produced by bacteria. In fact, Bt is not an obligate pathogen. Vegetative cells can be induced to sporulate and produce a crystal containing toxic proteins (? endo-toxins) when they are grown in sub-optimal regimes in fermentation chambers (Speight and Wainhouse 1989). The FSC has misgivings over the use of transgenically introduced Bt toxin, despite studies showing that Bt transgenic plants can often provide a pesticide delivery system that manages insect resistance better than pesticide delivered through Bt sprays (Roush, 1994). Whilst it is undecided if pest control using spray Bt as a whole bacterium is dangerous (Swadener 1994), it may be safer to use Bt transgenes incorporated by GM, and so reduce the probability of gene transfer between bacterium (Helgason et al. 2000).

Concern about the "build up of resistance to natural biological controls or the chemicals used to control them" (Soil Association 1998) is legitimate and of considerable importance. Pests acquiring resistance from elevated exposure to selective controls could be problematic for future generations. Insect resistance to Bt will eventually develop, GM or no GM, because Bt sprays have been used extensively by organic farmers since the 1950's (McHughen 2000). The same is true of chemical pesticides, used by conventional farmers for decades. In fact, given that the precision of delivery is increased with GM, there may be a lower likelihood of the development of insect resistance to Bt toxin with GM crops, especially if refugia of non-GM trees are used (Roush 1994).

The use of many chemicals is permitted under UKWAS, and thus FSC, guidelines (UKWAS 2000b). Some have potentially severe impacts, and it is these chemicals with which GM must be compared against. For example, permethrin is used for the control of Hylobius abietis, a weevil that can devastate restocked stands (UKWAS 2000b). Described as "low selectivity, extremely dangerous to aquatic life, dangerous to invertebrates and toxic to bees (do not apply when flowering vegetation present)." It "may damage non-target vegetation", and has been "identified as a potential endocrine disrupting chemical by IEH (UK Institute for Environmental Health) and EA (Environment Agency)" (UKWAS 2000b). Yet under UKWAS and UK FSC standards, use of permethrin is acceptable. This practice is contrary to FSC International standards (FSC 2000a), a position expounded by the Swedish FSC standard: "substances, including chemical pesticides and herbicides in the Chemicals Inspectorate's class 1 and 2 that are harmful to the environment and health shall not be used for the treatment of forest land. Permethrin treatment is currently exempted (until and including 1999)" (FSC 1997). Why would an exception be made for a chemical that otherwise contravenes FSC standards. It is the only means of controlling a potentially devastating pest. Other chemicals *e.g.* granular carbosulfan, and remedial options, *e.g.* weevil guards, are feasible but are more expensive . Moreover, "plants may require additional treatment(s) after planting, especially if small plant sizes are used" (UKWAS 2000b). Permethrin is certainly the cheapest, most commonly utilised and widely vocated practice. It may deemed acceptable because it has been used for years.

GM could conceivably provide a more 'ecologically friendly' alternative. Bt toxins are very specific insecticides, and target only susceptible species (Raffa et al. 1997; Speight and Wylie 2001). A Bt toxin specific to H. abietis could be inserted into the plant, just as has been done in Populus against the beetle Chrysomela tremulae (Moffat 1996). This would target only the pest, and possibly closely related species, which both predated the plant; a very precise application of insecticide. In contrast with the current practice of using permethrin, such an application of GM may be justified under DSS guidelines: "the use of selective pesticides may offer less impact than some non-chemical methods that are not species specific" (UKWAS 2000b).

Although not taking a specifically GM stance, 5.2.b of the PEFC standards states "inappropriate use of chemicals or other harmful substances or inappropriate silvicultural practices influencing water quality in any way should be avoided" (PEFC 1998). 6.3.3 of Annex 2 of PEFC Sweden is more succinct: "the use of chemical treatment is minimised and only used when other suitable methods do not exist" (PEFC 1999). In light of the earlier DSS discussion (pages 13-15), which describes appropriate methods of soil, weed and pest management, it is fair to comment that using chemicals can be justified in certified forests. One could thus argue for GMO certification following a rigorous assessment of impacts and risk. Interestingly, PEFC Sweden goes on to state "It is especially noted that permethrin is allowed during an exception period".

Whilst potentially reducing environmental impacts, transgenic plants producing pesticides do not allay the fears over increased pest resistance. The introduction of single genes may be the easiest approach to resistance, but single genes are susceptible to shifts in the gene pool following selective pressures, and are thus vulnerable to mutations within pest populations (Pimentel et al. 1989). This is particularly pertinent for tree crops, which have years of exposure in a single rotation. In agricultural crops monogenic resistance is overcome in a few years (Wilcox et al. 1996), so pests could devastate tree plantations in mid-rotation. This has been recognised by most proponents of GM, who recommend a very selective use of transgenic pest resistant trees until more genes of diverse function have been identified (Strauss 2000a). However, every gene system and not all simple resistance systems will be overwhelmed rapidly. For example, linkage analysis in loblolly pine (Pinus taeda) showed that resistance to the rust Cronartium quercuum was monogenic and has been successful despite more than 40 years exploitation as a commercial crop (Wilcox et al. 1996).

If exposure to a pesticide could be reduced, the selective pressure on target species might decrease, hence reducing the rate at which a species attains resistance. This could be achieved by linking the expression of inserted genes to inducible promoters. A level of predation could then be tolerated before the trees transgenic defences were activated (Strauss 1998). Furthermore, since many pests attack trees at a specific age (usually juvenile), it could be possible to link transgene expression to developmental stage. This would also reduce the exposure of pests to a toxin. The most prudent compromise seems to be the use of transgenic resistance as a supplement to existing quantitative resistance developed through conventional breeding (Burdon 1999). Thus, coupled with Quantitative Trait Loci (QTL) analysis using Marker Assisted Selection (MAS), GM could actually help ensure more lasting pest resistance (Burdon 1999).

FSC Criterion 6.6 notes "management systems shall promote the development and adoption of environmentally friendly non-chemical methods of pest management and strive to avoid the use of chemical pesticides". As noted in the DSS, judgements about pest management must be made by the forest manager and tailored to site specifics (UKWAS 2000b). Importantly, GM could be viewed as an effective non-chemical method. A major motivation for transgenic research is that GMOs are expected to have fewer ecological impacts than use of current chemical practices. English Nature recognise that the use of crops modified for insect tolerance may have potential benefits for farmland wildlife, particularly if their use results in "better targeted or lower impact of agrochemicals" (English Nature 2000). Thus a GM tree producing allopathic or insecticidal chemicals could fulfil criterion 6.6.

Lignin modification

Trees modified for lignin, and indeed other traits such as pest and herbicide resistance, will only be of advantage using short rotation crops such as poplar, eucalypts and some pines (Griffin 1996; Strauss 2000a). Indeed, Jonas Jacobsson of AssiDomän stated "There may be an economic advantage [of using GM] if you have a plantation with a short rotation" (3C Associates 2000).

The modification of lignin biosynthetic pathways could improve the suitability of trees for pulping (Pullman et al. 1998). This can be achieved by reducing lignin content or altering the biochemical composition of lignin; both mechanisms may reduce the need for chemical inputs to treat the pulp. For example, in transgenic tobacco the modification of lignin biosynthetic pathways reduced the Kappa value by 15% (Halpin et al. 1994), *i.e.* 15% less chemical bleaching was required to whiten the pulp. Enhanced lignin extractability has been reported with GM poplars (Baucher et al. 1998; Whetton et al. 1998). The potential benefits are substantial, with estimates suggesting that pine cad mutants could save $4 per tonne of paper produced by the Kraft process (Jeremy Brawner, pers. comm.). Indeed, a 1% reduction in timber lignin content would increase pulp yield by 1% - 1.5%, and could save the global paper industry billions of dollars (3C Associates 2000). Moreover, reduced inputs of energy, chlorine and other hazardous chemicals could lessen environmental impacts proportionately.

Booker and Sell (1998) suggest that lignin modification could compromise the structural integrity of trees and reduce their natural defences, whilst the FSC is concerned about "adaptation and pest resistance of trees, rate of decay of dead wood, and soil structure, biology or fertility, arising from use of GMOs with modified lignin chemistry" (FSC 1999a). However, it appears possible to modify the lignin content of trees without compromising plant viability (MacKay et al. 1997; Franke et al. 2000), and it is likely altered lignin composition will have very minor consequences (Dickson and Walker 1997).

Clone 7-56 of loblolly pine (Pinus taeda) is a naturally occurring mutant with a modified lignin biosynthetic pathway (MacKay et al. 1997; Ralph et al. 1997). Conventional tree breeders long ago identified 7-56's value as a parent in the production of trees with improved growth and pulping characteristics and thus incorporated it into their breeding programmes. Consequently, 7-56 progeny are planted on a massive scale throughout the southern states of the USA (Jeremy Brawner, pers. comm.). There have been no reported negative side effects of using this clone or its progeny, and no certification programmes have an issue with 7-56, having certified several thousand hectares (Jeremy Brawner, pers. comm.). MacKay et al. (1995) showed that mutation of the cinnamyl alcohol dehydrogenase gene (cad1-n) is responsible for 7-56's modified lignin. Plants that have an equivalent phenotype to the cad1-n mutant have been generated for tobacco and poplar (Halpin et al. 1994; Lapierre et al. 1999). Thus the same phenotypic effect can be derived by both conventional and GM methods. It seems paradoxical that a conventionally bred CAD null tree can be accepted and yet CAD-deficient mutants generated through GM cannot. They have the same phenotype and are thus very likely to have the same ecological impacts. This example illustrates a problem for certification with respect to the use of GM to alter lignin. That is, 7-56 homozygotes and CAD-deficient plants derived through GM are phenotypically equivalent, discriminated simply by method of production. Moreover, the FSC's concerns over lignin modification altering "rate of decay of dead wood, soil structure, biology and fertility" conflict with other FSC-accepted practices. Conifers, such as pines, and angiosperms, such as eucalypts, can be interchangeably planted on the same land. Their very different lignin and wood chemistry, as well as other chemical differences between these diverse taxa, may bring about ecological changes greater than anything possible via modification of lignin through GM per se. Moreover, the wood chemistry of exotic species is thought to effect the ecology of plantations far more than lignin modification of a native species. It is this that makes some believe the FSC has "no place bringing lignin modification as a rational concern" (Strauss 2000b) regarding the certification of GM trees.

Transgene escape

Widespread anxiety has been expressed over the possible contamination of wild populations by transgenic plants, which could compromise natural fitness or introduce some components of fitness that would increase a species weediness potential (Rogers and Parkes 1995; van Raamsdonk and Schouten 1997; Burdon 1999; Ellstrand et al. 1999). The FSC have detailed their concerns over transgene escape , and make particular reference to the potentially negative impacts arising from the spread of herbicide resistance .

Assessing transgene escape is very complex (Kjellsson 1996; Ammann 1999). In brief, considerations must be made regarding: (1) the likelihood of gene flow; (2) the probability of gene establishment (introgression) (Ellstrand et al. 1999); and (3) the impact of an introgressed transgene, which will depend on its effect on the ecological factors regulating the recipient plant population (Crawley et al. 1993; Dale 1994; Schmitt and Linder 1994; Gray and Raybould 1998). Assessing and quantifying these variables draws on theoretical ecology, comparisons with introduced exotic species and recently conducted experimental work. Trees present additional obstacles for analysis because of their long generation times, large size and potential for long distance dispersal of pollen and seed (DiFazio et al. 1998). Thus each transgene and its application must be examined in isolation. These factors have yet to be assessed, since not a single GM tree has yet received approval to reach sexual maturity in a field environment (Griffin 1996). Consideration is given to gene flow, and increased 'invasiveness'.

Gene flow

To date, most studies have focused on gene flow via pollen in agricultural crops (van Raamsdonk and Schouten 1997). This follows a rational call to assess each species independently due to differences in pollen biology: wind-pollinated plants will disperse their pollen over much greater distances than insect-pollinated species. Pollen may flow into the gene pool of a nearby species, by either outbreeding depression (fitness reduction following hybridisation) or genetic assimilation (dilution of the genetic integrity of the wild species until it is effectively assimilated into the crop species, Ellstrand 1992). This is not restricted to GM plants, but is a potential problem for any plantation, particularly if it constitutes a narrow gene pool since similar genes could 'flood' neighbouring populations. For example, hybridisation with non-transgenic crops has been implicated in the extinction of at least five wild relatives of food crops (Small 1984) and the Californian walnut (Juglans hindsii) (Ellstrand 1992) .

The more taxonomically distant an exotic tree is from native species, the lower the chance of gene transfer. Conceivably, if sufficiently distant exotic species of GM trees were used, there would be no risk of transgene passage to native plants. They could thus be certified. Conversely, the more closely related a crop tree is to indigenous wild relatives, the greater the likelihood of gene flow. Thus, in the UK, crops have been classified based on their potential to hybridise with wild species (Rogers and Parkes 1995); black poplar (Populus nigra) and Scotts pine (Pinus sylvestris) have been placed in the highest risk category. Physical proximity is also important; pollen dispersal experiments have confirmed a negative exponential or leptokurtic range of pollen distribution (Scheffer et al. 1993). However, gene flow in poplars has been demonstated over distances of at least 10 Km (Strauss et al. 2000c). This research exemplifies the potential for transgene spread; limited gene flow is inevitable if GM plants are grown close their relatives (van Raamsdonk and Schouten 1997; Ellstrand et al. 1999; McHughen 2000, p.166). It is also noted that "it is a clear mistake to place emphasis on pollen-based gene flow and nothing on the more obvious [sic. seed] route for gene escape" (McHughen 2000, p.166), for this enables gene flow through time as well as space. It is probable that legislation will require sterility in GM trees (Burdon 1999).

If genes do manage to pass between related species, several modifications are predicted to give plants a selective advantage, including enhanced tolerance to environmental factors, such as salinity, and pest resistance (Rogers and Parkes 1995). Herbicide resistance is only likely to confer a selective advantage to plants if they are exposed to herbicides. However, plants carrying these traits may be disadvantaged due to the metabolic costs of synthesising proteins in the absence of a selective advantage. Experiments with perennials generated mixed results, suggesting there are no costs associated with resistance genes (Lavigne et al. 1995) but weediness is not increased (Crawley et al. 1993; Snow et al. 1999). However, research is yet to be performed on trees, and is also needed to determine if there is a greater risk attributable to GM transgene spread than to conventional gene spread. However, one of the problems facing regulators and users of GM plants is the quantification of small risks from such experiments, and then scaling them up to plantation scale (Rogers and Parkes 1995). Large-scale releases of pollen will result in a larger flow of genes to wild relatives. Such genes may increase in frequency in the wild population irrespective of their fitness effects, such that a large scale plantations may be sufficient to overcome any deleterious fitness effects of the transgenes (Dale 1994; Gliddon 1994; van Raamsdonk and Schouten 1997). This issue is also not unique to GM plantations, but to any large scale plantation.

The risk assessment of gene escape also depends on the nature of the gene. Again, this is equally true of conventional and GM plants. For example, in agriculture 'Smart Canola' has been conventionally bred to express resistance to two herbicides (Concar 1999). Such plants present very similar risks as GM ones, and should therefore be subject to the same risk assessments, for "if the concern over gene spread is valid, the method of initial production is irrelevant" (McHughen 2000, p.113). Thus, the certification requirements imposed on any plantations should be examined.

Most certification systems make specific reference to utilising appropriate provenances, varieties and species . These criteria, particularly those of the PEFC, are designed to address principally the concerns over gene spread between endemic species. Some criteria are very specific, for example 6.3 209 FSC UK requires that "species diversity is maintained and dilution of the local gene pool is minimised" (FSC 1999b). Indeed, FSC UK set stringent requirements on total species composition. However, the caveat 'appropriate' and similar dispensations appear in most of these criteria: "Seed of local provenance is used wherever it is available and considered appropriate for planting and restocking of native species" (9.2 335 FSC UK); "for reforestation and afforestation, origins of native species and local provenances that are well adapted to site conditions should be preferred, where appropriate." (4.2b, PEFC 1998). These exemptions recognise that sustainable forestry is a balance between ecological and financial objectives, and ultimately dependent on numerous variables for which it would be impossible to specifically legislate.

Certification may be awarded for planned improvements, and forest managers may make extremely literal interpretations of the criteria. Whilst the criteria aim to prevent subjectivity, ultimately the interpretations of individual assessors will have a role, which may explain why some certification standards have been criticised (Counsell 1999).

Increased 'invasiveness'

"Considering the analogy with exotics has been a helpful one in regulating the release of GM" (Nuffield Council on Bioethics 1999); this is because the community level effects of transgenes acquisition by wild relatives have been hypothesised to be similar to those of introduced species (Schmitt and Linder 1994; Williamson 1994; Rogers and Parkes 1995; Rissler and Mellon 1996). Most seriously, the competitive superiority of GM plants, as a consequence of an acquired transgene, may lead to the exclusion and extinction of the native plants (van Raamsdonk and Schouten 1997); "it is the traits that increase competitive behaviour that are of primary concern" (Nuffield Council on Bioethics 1999). The effects of increased competitiveness can cascade through the ecosystem (Rissler and Mellon 1996), for example, in the United States, 42% of the species on the threatened or endangered species list are at risk primarily because of non-indigenous species (USDA 1999) costing the US economy an estimated $138 billion a year (Pimentel et al. 1999). However, some prominent GM scientists "do not believe that transgenics have properties even remotely similar to invasive exotics, and should not be considered along with them for scientific or regulatory purposes". Exotics form the mainstay of industrial plantations in many countries, and total 98% of the US food system (Pimentel et al. 1999). This is because they grow so well; 10.4, 363 FSC UK notes "...native species are preferred. Exotic species are only used where they will substantially out perform native species in terns of meeting the objectives of plantations" (FSC 1999b).

In New Zealand 6% of the country is covered in plantation forest. This generates the majority of US$1.42 billion of timber product exports (FAO 2000b), which accounts for 5.5% of national export earnings (ODCI 1999) . Most of the plantations are of the exotic Pinus radiata, which has been intensively selected, converting the poor wild form into a straight-trunked, fast-growing tree crop (Chilvers and Burdon 1983). Pinus radiata can be highly invasive (Cronk and Fuller 1995), and is considered to be a 'significant problem' in scrubland, forest margins, sand dunes, open land and short and tall tussockland (Williams and Timmins 1990). Fletcher Challenge Forests recently decided to opt for FSC and ISO certification of their New Zealand forests, in order to maintain their position in U.S. markets (Kelly 2000). Earlier FSC certifications in New Zealand, such as the Rayonier's 34,000 ha Southlands Estate, have not considered exotic species inherently problematic. As FSC 6.9 states, "The use of exotic species shall be carefully controlled and actively monitored to avoid adverse ecological impacts". The situation is similar in Scotland. 6.1 173 FSC UK states "operations planned with consideration of: the spread of invasive species across the forest boundary in either direction". Thus, it could be argued that the certification of exotic and/or invasive species is possible when contingencies exist for potential negative impacts.

The issue of invasive exotics is also pertinent in South Africa, where timber and pulp production are almost entirely dependent on exotic species – pines, eucalypts and Australian acacias ('wattle') (von Maltitz 2000). Many of these species have been demonstrated to be highly invasive, particularly in the Fynbos

(Cronk and Fuller 1995) and veld-grasslands (Cooper 1999). Indeed, SAPPI have a special eradication programme for alien invasives. Yet three forestry companies have hundreds of thousands of hectares of FSC certified exotic, potentially invasive, forest (FSC 2000c; Von Maltitz 2000). This has drawn criticism from environmental groups in South Africa such as WESSA, who have expressed concern about the credibility of the FSC over their policy surrounding exotics (Murphy 1999).

If the potential impact of GM trees is to be analysed, the risks must be considered against current policies for exotic species. The introduction of non-native species is specifically warranted by the FSC. For example, the great spruce bark beetle Dendroctonus micans is a pest of exotic spruce plantations (Speight and Wainhouse 1989) and costly crop losses can be controlled by the exotic predatory beetle Rhizophagus grandis. Thus, 6.9 270 FSC UK states the "use of non-native biological controls such as Rhizophagus grandis may be desirable to control non-native pests". The use of this exotic beetle reduces economic costs and insecticide application. Its interactions with the environment have, and never will be, fully elucidated, but the benefits are deemed greater than the potential impacts. These arguments ring true for GM.

RESTRICTED ACCESS TO ADVANTAGES

The development costs of GM trees is great. Thus, even large forestry and biotechnology companies are collaborating in projects such as ArborGen . However, investing millions of dollars in this field is risky, and only deemed acceptable since patent protection provides a temporary monopoly over new products, enabling companies with the leading technology to dominate their sector. This has raised concern amongst NGOs, who fear poorer growers will be unable to afford increased capital expenditure, and thus have restricted access to the advantages of GM. The FSC acknowledge this anxiety . However, justifications for sharing of intellectual property will be very difficult, and out-of-line with the increasing harmonisation of global patent protection.

However, those pursuing GM technology argue that the higher cost of GM-trees is a demonstration that the trees have additional value; growers will only purchase new products if the additional value is worth the higher price. For example, a traditionally bred hybrid maize seed is purchased by most U.S. farmers since it generates more income, justifying the extra initial expenditure (McHughen 2000, p.192). This is despite having to buy new seed each season. In contrast, the GM tomato, Flavr-SavrTM, failed because it was too expensive and the flavour did not warrant consumers spending the extra money (McHughen 2000, p.257-258). However, many small growers, especially in developing countries, cannot afford capital outlay. As illustrated by the Green revolution, additional costs favour larger growers. However, this issue is not specific to GM, and it would be wrong to consider GM in isolation.

If we consider current practice, many growers already purchase their seed or seedlings from specialist growers; in vitro generated clones and advanced breeding programmes are beyond the scale of all but large forestry companies or specialist nurseries. If foresters wish to plant improved varieties, invariably they buy from these suppliers, since their access to improved varieties is already restricted, regardless of whether the product is GM. Although certification schemes make specific references to using local provenances where appropriate (PEFC 1998 4.2b; FSC 2000a 9.2), the onus is on genetic diversity. Reference is not made to restrictive or monopolised access to quality seed.

Certification is espoused as a means of ensuring socially responsible and sustainable forestry. Yet critics have noted that becoming certified imposes additional costs upon growers (Centero 1998). This additional expenditure, particularly that of the FSC system, is seen by some to exclude small growers (Counsell 1996), especially in developing countries (Banahene 2000). In a market increasingly concerned with a reliable supply of a standardised product, certification makes it even more difficult for small growers to gain access to markets (Counsell 1996). Some have considered a major driving force behind certification to be large companies intent on expanding their market share (Counsell 1996). This may explain why 66% of forests certified under FSC guidelines have been by large industrial enterprises (Thornber 1999). Although the certifiers have acknowledged the difficulties faced by small growers, and have tried to develop group systems, the additional expense and limited supply of certifications is restrictive.

Reduced biodiversity from sterile trees

Due to the potentially negative effects of gene flow, sterility may be required for the release of GM trees (English Nature 2000). The ecological impacts of large sterile plantations has raised concerns (Dr George McGavin quoted in Tickell and Clover 2000) over reduced biodiversity. The FSC is specifically concerned about the "Reduced biodiversity of organisms dependent on flowers and fruits, arising from use of sterile GMOs" (FSC 1999a). Moreover, the FSC has doubts over the ability of trees engineered for sterility to prevent gene flow to native plants .

Engineered sterility is seen mainly as a risk reduction measure to minimise negative effects of transgenc flow and genetic erosion. However, sterility can offer other benefits. Reproductive growth requires between 15% and 30% of a trees energy (Owusu 1999). Thus, suppression of reproductive tissues could channel more resources into vegetative growth and thus could greatly increase productivity. Furthermore, it could reduce airborne allergens.

Sterility can be conferred in two ways. The first involves the suppression of the floral genes that are essential to produce fertile gametes. This can be achieved through either antisense suppression of gene expression or homology-dependent gene silencing . The principal disadvantage of these approaches is that native genes or highly homologous equivalents are needed. This greatly increases the research time and expense of modification. Furthermore, floral development is very complex, and thus redundancy would have to be built into the system in order to ensure stability throughout a plant's life, and would require many genes to be used. For example, LEAFY (LFY) in Arabidopsis controls a developmental switch to convert lateral shoots into flowers (Weigel and Nilsson 1995). The main stem must acquire competence to respond to LFY, and this capability increases during the life cycle (Weigel and Nilsson 1995). Thus, genes prior to LFY control LFY expression, and include promotors of flowering such as COSTANS and inhibitors of flowering such as TERMINAL FLOWER. Research using PTFL, a poplar LFY homologue, indicates that these promotors and inhibitors are in turn governed by a complex system of genes that respond to day-length (Rottmann et al. 2000). Moreover, genes downstream of LFY control precise floral organ development (Coen and Meyerowitz 1991). By targeting different genes in this complex system redundancy may be ensured.

Suppression of floral genes is attractive for modification since fertility is impaired at various stages of floral differentiation (Melian and Strauss 1997), which enables selective sterility to be generated in tissue-specific regions. For example, modifying structural or catalytic proteins essential for pollen formation could generate male sterility. Whilst safe-guarding against gene flow via pollen, breeders could still cross female flowers with pollen producing trees. However, such modifications of pollen producing genes may not enhance vegetative growth, because expression is late in floral development. Moreover, the problems of seed production and gene flow via seed still exist, and the FSC is specifically concerned about partial sterility that does not entirely prevent gene flow (FSC 1999a).

The second means of generating sterile trees would be through the expression of cytotoxin genes in a floral tissue-specific fashion. This would disrupt or ablate organ-specific tissues and since early and frequent floral-specific expression, floral homeotic gene promotors are probably best suited for engineering complete sterility, whilst enhancing growth (Meilan and Strauss 1997). However, occurrences of instability in transgene expression have been reported (Denis et al. 1993), and it is these instances that concern ecologists and the FSC about transgene spread. Trees remain in situ for many years, and are seldom closely monitored. A breakdown of sterility could go unnoticed for many years and allow GM genes to escape to wild populations.

If sterility is an important aspect of gene containment for regulatory and certificatory purposes, it is critical to demonstrate that sterility is maintained under a range of conditions and development stages. This requires non-contained field trials. It will take many more years of work to develop guaranteed sterility systems (Strauss 2000a). Until then, existing sterility systems will be a risk reduction measure, and require an acceptance of small transgene releases. However, the problems of time-scale, massive pollen releases and the rarity of the breakdown in sterility makes monitoring difficult and costly. Coping with a breakdown in sterility and the potential consequences of transgene releases could be impossible. However, equivalence must be invoked when considering the ecological aspects of sterile plantations. If the technology is regulated properly and used with consideration, there is no reason to expect a negative consequence of failed sterility transgenes. Furthermore, it is unlikely the consequences of failed sterility will be as dramatic as those described for Pinus radiata introduction (Cronk and Fuller 1995).

Currently, large plantations of exotics are certified. The productive areas of these forests are often highly effect environments, with a biological composition radically different to the wild systems they replace (Peterken 1992). Certification systems acknowledge the detrimental impacts of exotic species, but agree the increased yield from non-natives justifies their use. Thus they aim to ensure ecosystem management to maximise beneficial activities and minimise adverse impacts. Equally, these ecosystem management practices could be applied to sterile trees, since the potential for increased wood production is great and certification systems could justify greater proportions of the forest dedicated to ecologically beneficial practices. GM is likely to be used with exotic species in short-rotation, commercial plantations (Griffin 1996; 3C Associates 2000; Strauss 2000a). The ablation of reproductive organs, and consequently the loss of reproductive food matter to organisms in these plantations, would represent only a small impact when contrasted to those already imposed through the use of exotics. The very reason exotics grow so prolifically is thought to be their unpalatablity to a significant number of pests (Speight 1989).

General concerns

The FSCs concern over "Reduced adaptability to environmental stresses, changes to interaction with other organisms, and increased weediness or invasiveness, in GMO trees with new features" (FSC 1999a) encapsulates all their previous points. Reference to the preceding sections should be made to discuss these specific issues. However, treating such issues as disparate overlooks the potential problem that these factors may compound and interact. The resultant gestalt could be unpredictable and pose potential problems.

EVALUATING THE RISKS

PREDICTABILITY AND INSTABILITY

Although not expressly voiced by certification systems, the unpredictability of genetic transformation has raised concerns in the scientific community. This is because the "Insertion of a novel gene can have a collateral impact on the rest of a hosts genome, resulting in unintended side effects" (Owusu 1999). New genes that could conceivably disrupt the metabolism of an organism will usually produce transgenics that are lethal. However, experience with maize showed that a cytoplasmic male-sterility factor led to the breakdown of resistance to the rust fungus Bipolaris maydis (Levings 1990). Although the technology used in this example is dated, genetic instability could produce many unintended effects that only manifest themselves years after deployment. The vast majority of transgenes are stable but there have been instances of instability, including altered patterns of gene expression (van der Hoeven et al. 1992) and the failure of engineered sterility (Denis et al. 1993).

Instability may be induced by high temperatures (Broer et al. 1992; Meyer et al. 1992; Walter et al. 1992), which are thought to alter methylation (Meyer et al. 1992). This illustrates an important principle. Since many genes are environmentally and developmentally regulated, the physiological state of plants can have an important effect on transgene expression, thus, the risks of collateral alterations to stress-activated genes "cannot be anticipated until the stress response is actually triggered" (Owusu 1999). This is of special import for forest trees because they are exposed to environmental fluctuations for much longer periods than agricultural crops. Moreover, a time lag could make the problems of instability more acute because the widespread planting of a particular cultivar could be spaced over many years. A greater time frame also has a bearing on the uncertain risk of novel gene mutation.

However, some have pointed out that in 1999 alone GM crops covered 45 million hectares (100 million acres) in North America, and there are no documented dramatic or adverse or unexpected effects from any of these plants (McHughen 2000, p.190). Studies with poplars have shown stable gene expression for several years (Strauss, pers. comm., 2000). The rigorous testing and regulatory policies prior to commercial GMO release are considered effectively to remove unstable cultivars. Extensive quantities of data are gathered in trails, and the quantified risks satisfy assessors and government legislators, although they may not allay the concerns of some groups who question the independence of assessors. Simple steps could also be taken to lessen the potential pitfalls of instability. Just as for risk aversion in clonal plantations, a variety of transformants could be planted. These could have a number of transgenic insertions with multiple copies of the desired transgenes (Burdon 1999). Again the debate centres on risk analysis, and as such should focus on the specific GM trait.

It is important to note that "all living things are subject to natural genetic instability" (McHughen 2000, p.189). Pieces of DNA called transposable elements can move directly from one chromosome site to another (Fedoroff 1992), multiply, and occasionally re-arrange neighbouring DNA sequences

(Alberts et al. 1989). They are thought to make up 10% of higher eukaryote genomes, and can affect gene regulation (Alberts et al. 1989). Such inherent instability is in present in all plants. Genetic changes induced by transposable elements are accommodated by phenotypic plasticity. As such, GM would not present any new phenomenon of instability.

Risk assessment

Some FSC supporters are against the deployment of GMOs (Greenpeace 2000), whilst other supporters call for a moratorium (Owusu 1999). The spectrum of potential GM applications has lead many bodies to recommend that "regulators should explore the pros and cons of adopting a more explicit risk / benefit assessment" (Nuffield Council on Bioethics, 1999). Risk assessment can be controversial, reflecting the important role that both science and judgement play in drawing conclusions about the likelihood of effects on human well-being and the environment; contention often arises from incomplete knowledge. To make an effective risk management decision, stakeholders need to know what potential harm a situation poses and how great is the likelihood that this harm will be realised. Each application should be examined in isolation; however, the hazards of GM cannot be properly evaluated (Apel 2000). While the extent of exposure can be estimated, there are no responses to exposure. Contemporary risk models must thus make assumptions (Burdon 1999), and herein lie the seeds of dispute. Those against GM extol the precautionary principle based on our lack of knowledge, whilst others maintain negative assertions can never be proven, and GMO deployment should be based on comparisons with current practices. This has lead some to argue that the "anti-biotechnology contingent has made the lack of scientific knowledge part of a strongly subjective standard of risk assessment, and coupled it with a confirmation bias " (Apel 2000). Countering this, some of the scientific community believe "... the simple principle of genetic modification spells ecological disaster. There are no ways of quantifying the risks... ...The solution is simply to ban the use of genetic modification." (Narang 2000). This dichotomy of views may rest with philosophical beliefs surrounding 'natural' procedures and the role of multi-nationals in dictating much of biotechnology research in forestry. As such societal interests play a strong role in moulding the perceptions of GM. Reconciling these different perspectives may be very difficult.

Proving GM is non-hazardous requires a negative proof, and science cannot corroborate such assertions. A logical approach is to compare the risks of the specific product in question with currently acceptable risks. However, the notion of substantive equivalence is not often popular with official bodies because it may expose existing practices as flawed. Yet equivalence is the rational scientific approach (McHughen 2000). As such, an examination of currently acceptable practice for certified forests should reveal comparable requirements for the certification of GM forests. One of the greatest justifications of anti-GM lobbyists are the potential impacts of GM. Many thus advocate the precautionary principle, at least until the variables are known and a risk assessment can be completed. Yet, consider PEFC prescriptions: "acidification could be controlled by liming. The short term and long term ecological impacts of large scale liming are not fully evaluated" (C5.3, PEFC Sweden, Annex 4, 1999). Liming to combat acidification is considered acceptable, despite its impacts not being fully evaluated. This is contrary to the precautionary principle. Many could draw a parallel between accepted practices such as these and the certification of GM. Indeed, practices considered by some to be very damaging to the environment are justified by increased harvest; "the removal of tops and branches and rotten round wood as wood energy is a supplementary harvest. To compensate for nutrient loss, ashes from wood burning shall be brought back to the forest or compensating fertilising shall be done according to special rules" (Criterion 3:2, PEFC Sweden Annex 5, 1999). Again a value judgement of cost and benefit has been made after assessing potential risks. This would argue that GM applications could be examined on a case by case basis, rather than imposing an outright ban. That is, "focus on the risks of the product, not the process", (McHughen 2000, p.159) just as in conventional forestry.

Since the application of GM cannot be analysed by conventional risk assessment approaches, alternative methods must be carried out. This should be based on scientific fact and substantive equivalence. Because the philosophical bent of individuals is relevant, assessors should be "cognisant and transparent ", combining social awareness with the tenets of rigorous science (Farnham et al. 2000). This ensures a more realistic elucidation and quantification of the trade-offs faced. In order to ensure such deductive distinctions are as objective as possible, knowledge should be integrated through discussion and debate. Nevertheless, it is important to note that "in every study so far, no evidence has been found that GM crops present special risks. The types of risk are exactly the same as for crops modified by the classical plant-breeding methods" (Cook 2000).

If certification systems are to accept GMOs, they have two options. They could accept the judgement of the national systems that currently assess applications for GMO release. This would be very simple, but the differences between national systems could be contentious; some nations are seen as having more rigorous standards than others. However, if certifiers are not satisfied with the stringency of a national system they could impose requirements to meet their own, or another countries, more demanding criteria. Given that it is likely GMO plantations will first appear in developing countries (Owusu 1999), and that national legislation and its implementation may be lacking, it is possible that certification represents the only means of ensuring transgenic releases are safe. As such, it is important for certifiers to recognise that despite never being able to quantify precisely the risks of GM, it is important for them to examine their policies and develop a more sophisticated and realistic approach to the certification of GMOs. Moreover, since it has been shown that silencing of one transgene by another may occur (Matzke and Matzke 1991), widespread monitoring of GMOs has been proposed (Rogers and Parkes 1995). Certification systems could be in a unique position to ensure this monitoring is carried out on a large scale.

As always, it is the application and specifics of the GMO which should be assessed. Risk analysis is very difficult since all variables cannot be known. If GM is to be accepted this will require a flexible certification system. In light of the current permitted practices, from a rational perspective, many applications of GM could be certified, even without risk assessment. Having said this, it is important to note that this is considering GM applications on a case-by-case basis.

CULTURAL ASPECTS OF FORESTS

Forests play a vital role in the life and culture of people around the world. The reverence and adoration of trees has a strong psychological and social foundation in most human cultures. The variety of cultural values and symbolic functions ascribed to forests are as numerous and diverse as the communities and cultures, Forests feature in all aspects of culture: language, history, art, religion, medicine, politics and even social structure.

In many African cultures trees feature in myths and lores. Forest trees serve as link between the sky and earth and is associated with creation as well as the underworld. In parts of West Africa forests provide the venue for many cultural events. The 'arbre a palabre' are locations for social and political meetings. The locations where elders sit under big trees and talk, argue and discuss issues until they agree. It is a place where political, social and judicial decisions are made. In Cote d'Ivoire tree species such as Blighia sapida, Cordia millenii and Bombax buonopozensa are some of the preferred tree species. The Oubangui tribe of Central Africa, plant a tree for every new born child. For female children a fast growing tree species is planted. The child development is linked to the tree growth. If the tree growth declines there is fear for the health of the child and a healer is called upon. When the child is sick she is brought to the tree for ritual treatment. When the tree starts to fruit, the time would have come for the child to marry. When a person dies his/her spirit is believed to go and reside in his/her personal 'birthright' tree.

In other regions of the world also, there exist a relationship between forests and spiritual realm. Budda would sit alone in the depths of the forest in meditation and it was in the midst of a forest that He was shown the four great truths. The Dai people of Yunan province in China believe that the forest is the cradle of human life and that forests are at one with the supernatural realm. The forests in European culture were also considered to be more positive sites of miracles, the search of great spiritual awakenings and the forest itself was held to be form of primitive church or temple. The first temples in Europe were forest groves. Ties to nature manifest themselves most notably in Turkish culture. After conversion to Islam, the importance of trees grew in Turkish culture because Mohammed compared a good Muslim to a palm tree and declared that planting a tree would be accepted as substitute for alms.

SACRED GROVES

In sacred groves are manifested a range of traditions and cultural values of forest throughout the world. Sacred groves are specific forest sites imbued with powers beyond those of human. They are often sites of ancestral burial where people can communicate with their ancestors. Trees within these groves are considered as sacred, housing spirits and providing links to ancestors. In some areas sacred groves are the only intact forests remaining. Although many cultural traditions are disappearing with rapidly changing social and physical environment, sacred groves often remain as valued elements of cultural heritage. The groves are also often the site of ritual healings and location where villagers find particular plant medicine.

Access to most sacred forests are restricted by taboos, codes and custom to particular activities and members of community. Gathering, hunting, woodcutting and farming are strictly prohibited to the Dai people in the holy hills of China. However, control over extractive activities in sacred groves varies in communities and cultures.

In some communities and cultures a complete ban is not in place, and limited collection of fallen wood, fruits from forest floor, medicinal plant collection, honey collection and other activities are permitted, if strictly controlled. Sacred groves in general have survived for many hundred of years and today act as reservoir of biodiversity or library of nature.

In some areas sacred places play a major part in safeguarding critical sites in the hydrological cycle of watershed areas. In different cultures some specific forest resources are revered or serve as religious or cultural symbols. The birch in Scandinavia, larch in Siberia, redwood in California, Fig in India and Iroko in West Africa are widely revered and respected. The oak tree was worshipped by Romans, Druids, Greeks and Celts as the home deities. In Europe fairies were said to have made their homes in old oak trees, departing through holes where branches had fallen; it was considered healing to touch the fairy doors with diseased parts.

Ceiba pentandra (associated with burials and ancestors in the Amazon), Copaifera religiose (associated with fecundity, wealth, power and fame in the

South America) and Milicia excelsa (associated with fertility and birth in West Africa) are sacred to the people. Some trees also serve practical judicial roles. Practically they are physical boundary markers that define property and provide evidence of usuary rights in judicial disputes. In many traditions in Senegal and Cote d'Ivoire trees play a central role in the land tenure system. Planted trees evidence land use rights for individual or lineage groups. In Ghana there are court cases that have adjudicated in favour of individuals who have planted and tended naturally regenerating fruit trees on a piece of land for several years without interference as constituting a proof of possession.

TRADITIONAL KNOWLEDGE AND SUSTAINABLE FOREST MANAGEMENT

It is estimated that over 300 million people live in or near forests and obtain part or all of their livelihood and food from forests. Forest is key to non-farm employment for the forest fringe communities. Over the years they have used harvesting methods that are ecologically benign which have ensured their sustenance and survival.

The paradigon of sustainable forest management (SFM) has been widely embraced in national and international policy levels, but it has not yet been implemented to the point where it is appreciably mitigating the negative trends affecting the world forests particularly in the tropics. SFM provides increasingly sophisticated set of policies and tools for managing forests in a more sustainable way. Implementing SFM, however, requires overcoming many of the same economic, political and institutional hurdles that drive deforestation and forest degradation.

Forest management would benefit from incorporation of traditional knowledge of indigenous and local people. Traditional forest related knowledge (TFRK) has long been known to have implications for forest management, conservation of forest biodiversity and identification of forest genetic resources. Traditional knowledge and practices have sustained the livelihoods, cultures and forest resources of local and indigenous communities for centuries. This knowledge most often inter-woven with traditional religious beliefs, customs, folklore, land use practices and community level decision making processes, has historically been dynamic, responding to changing environment, social, economic and political conditions to ensure that forests continue to provide tangible (food, fodder, medicine, water, soil) and non-tangible (spiritual, social, psychological health) benefits to the present and the future generations.

The current limitations of modern science to deal effectively with environmental issues of increasing magnitude and complexity, (*e.g.* climate change, biodiversity conservation) has opened the door for incorporating other sources of knowledge. Society can learn a great deal from the traditional skills in sustainably managing complex ecological systems. Traditional knowledge holders have developed extensive knowledge about the spatial and temporal distributions of natural resources and behaviours of many natural species and factors that influence them. This knowledge that is distinctive to communities, tribes, cultures arises from personal experience and practices passed on from generation to generation. Over the years local and indigenous people in Africa, Asia and South America have developed a variety of vegetation management practices.

TRADITIONAL KNOWLEDGE AND TRADITIONAL MEDICINE

Traditional medicine refers to health practices, approaches, knowledge and beliefs incorporating plant and animal based medium, spiritual therapies and combination to diagnose, treat and prevent illnesses or maintain well being. In Africa, Asia and Latin America, it is estimated that over 70 percent of the population use traditional medicine to meet their primary health care needs. WHO estimates that in Europe, North America and other industrialized regions over 50 percent of the population have used complementary or alternative medicine at least once and the global market of herbal medicine currently stand over $60 billion and growing steadily. About 25 percent of modern medicines are made from plants and many of these plant based modern medicines are used for the same purposes for which the local and indigenous communities used them. Indeed traditional knowledge underpins the traditional medicine practices.

The construction of traditional knowledge databases and knowledge archives about native groups uses of local plants is a sure way of combating biopiracy, blocking patents by multinational companies and ensuring that holders and users of traditional knowledge on traditional medicinal plants receive fair reward for such knowledge. Database of traditional formulations is needed to allow examiners to compare patent applications with existing traditional knowledge. Indeed, India is one of the countries that has successfully documented traditional knowledge and successfully fought patents granted at the USPTO on turmeric and at EPO on neem (Azadirachta indica). A number of countries have national regulations or in the process of preparing regulations on herb and medicines but the legislative control of medicinal plants has not evolved around structured model. It is proposed that an effective legal system for protection of intellectual property rights of traditional knowledge is established by governments to protect the traditional knowledge of local and indigenous people.

Traditional Knowledge and Climate Change

There are still major gaps in climate science. Traditional knowledge can inform and provide valuable insight into scientific investigations on the effects, impacts and coping strategies for climate change. Local and indigenous people have lived in changing climate over hundreds of years. As a result of their close relationship and dependence on forests they have developed and use diverse tools to assess the impact of changing climate on their communities and ecosystems and have developed and implemented resilience and adaptive strategies. Local observations of direct effects of climate change corroborates scientific predictions; and include temperature and precipitation changes; coastal erosion; changes in wildlife, pests and water borne diseases distribution; extreme weather events (*e.g.* drought, flood) and changing weather patterns. In addition many local and indigenous communities are able to forecast impending weather based on early warning signs (typically related to the sky and sea, movements of the sun and moon, changes in plant phenology, and changes in animal Behaviour).

For example there is already a vast store of information at the community level on monsoon prediction. Traditional knowledge can provide scientists and resource managers a long term perspective that is lacking in more typical observations and can establish the relations of historical impacts of land use and climate change. In the context of climate change research local and indigenous communities are providing important source of climate history and baseline data and local scale expertise. Traditional knowledge on climate change does not only augment scientific information but also promote scientific enquiry.

Integration of Traditional Knowledge and Formal Science

Despite their important contribution to SFM and sustainable livelihoods, TFRK and practices are fast disappearing. The negative implication of this loss of TFRK on livelihoods, culture and biodiversity and the capacity of the forest to provide the goods and services remain poorly understood, unappreciated and undervalued by policy makers and the general public in many countries. The question is whether there is any possibility of integrating formal science and ethno-science. Empirical evidence suggests the affirmative. Traditional knowledge systems indeed complement scientific knowledge system by providing practical experience in living within ecosystem and responding to ecosystem change. Traditional knowledge systems and scientific knowledge systems are not exclusive to each other and both are required to achieve SFM. For example a study in Nepal showed that Traditional Knowledge on firewood and fodder values corresponded to scientific assessment. The research revealed that local peoples preference of 16 firewood plants and 23 fodder plants were significantly related to firewood value index (FV1) and fodder value index (FoVi). In vegetation classification and distribution, management and growth characteristics traditional knowledge systems and scientific knowledge systems are in accord with each other.

TREES, WOODS, CULTURE AND IMAGINATION

We have given ourselves a conundrum, without answer. The very notion of creating new forests flies in the face of all we know, and much of what we practice. We know their great age, in the geological sense and domestically. We carved ourselves out of the forest in Britain and in our previous and elsewhere lifetimes. And ever since, we have spent our lives in clearings, on farms, in gardens, in parks, by the wayside, even on buildings, trying to stop trees growing. In parallel we have developed a working relationsliip with trees which has given us great cultural knowledge, affection and wisdom.

We cannot create what nature has spent millions of years learning to do. Forests of a single century demonstrate poverty .... but that is our startiug point, so we must seek and find the ancient and the old. We must protect the 'over mature' and ancient trees, and the nearest we have to ancient woodlands. We need the layers of before, aud if we cannot have it in the trees then it must be revealed in other things.

Beyond these let us relearn about succession ... that much underused branch of ecological insight ... let us watch the course of things. The way that one group of plants and animals will succeed another gradually working to an equilibrium which then stabilises and perpetuates itself dynamically. The gradual supplanting of seral stages moves at a pace which we can appreciate and accommodate. The gentle expansion of those wonderful May trees of the undergrazed upland edge and of greenbelt fields, the old pit tips covered in big trees now but first clothed in scrub. Why have we forgotten how quickly the birch grows and how fast its wood rots ... the perfect pioneer, the beautiful invader.

Trees are as much cultural beings as they are botanical specimens or timber. The tree is a powerful symbolic creature, as well as being a mysterious and miraculous one, hence its use and reuse to help us explain ourselves in the world (The Tree of Knowledge of Good and Evil, the Bo Tree, the Tree of Life). The deciduous tree demonstrates for us the paradox of death and rebirth every year, it embodies longevity, interconnectedness, generosity, hope and also darkness and fear. The forest carries for us many levels of meaning, for example in the boreal fairy stories (Little Red Riding Hood, Hansel and Gretel, The Two Brothers). Bettelheim (1976) in analysing the latter explains:

'The forests, where they go to decide that they want to have a life of their own, symbolizes the place in which inner darkness is confronted and worked through; where uncertainty is resolved about who one is; and where one begins to understand who one wants to be.' Perhaps we have a responsibility like the fairy stories to create for ourselves forests in which we can test ourselves, into which we can read answers which beckon us on to deeper understandings. Forests we can get lost in, safely. Archaic societies have a sophisticated understanding and systems to deal with these things, we flounder around pretending that because they are not economically graspable, not mathematically cageable, not financially fixable that they are not important.

In creating new forests we have to remain sensitive to what already exists, landscapes we may consider bland and those most changed by our activities, carry with them much of significance for those whose life and work has revolved around them. Nowhere should be regarded as a site for rehabilitation, before considered debate with local people. An important leap of the imagination lies in recognising that local knowledge and feeling have much to offer in the process of change. That involvement of local people, as individuals and in groups can offer a rich plural system for planning and energising the forest. Parish Maps focus on the less tangible feelings which people have about their places, starting with the subjective question 'what do you value here?' This makes them the expert, and provides a vehicle for the demonstration of knowledge, emotional attachment and ideas for the future, in a social, practical and creative way.

Trees have been watched, grown, revered, used, eulogised and understood for far longer than the rummaging of scientific arboriculture and the fashions of forestry. We need to know how to interweave new knowledge with the vestiges of the old: the young abstract knowledge, ebullient and often market driven, coining in from the outside looking for a circumstance to practice on, with the old wisdom in place, diffident, severed, unpractised.

In making new forests we must aspire to adding value to the patina of place, conspire to add new layers of significance, to mix in new meaning and detail. Not for the sake of it, but because places without meaning have no long term future and no short term richness. The 21st century forest has to have both. New forests mean new identity for the place, we would assert here the need to strive towards discovering and reinforcing Local Distinctiveness.

Trees have their territories, ecologically defensible but also culturally important. They have helped us to orientate and locate ourselves, trees that are comfortable in their place have added importance to us. D H Lawrence (1928) expressed it in talking of Eastwood, the country of his heart 'To me it seemed, and still seems, an extremely beautiful countryside, just between the red sandstone and the oak trees of Nottinghamshire, and the cold limestone, the ash trees, the stone fences of Derbyshire'; and Oliver Rackham (1986).

'Every oak or alder planted in Cambridge (traditionally a city of willows, ashes, elms and cherry plums) erodes the difference between Cambridge and other places. Part of the value of the native lime tree lies in the meaning embodied in its natural distribution; it is devalued by being made into a universal tree.' Landmark trees remind us of the hard work of the drovers finding their way across country (exotic trees have their place), the parish boundary or the sacred. And ancient trees guarded and venerated, what they are and where they are having melted into reinforcing significance, like the yew tree in the churchyard, or the oak in the centre of the village.

How can the process of creating a new forest 'go with the grain', highlight and accentuate what makes this place different and rich, create novel layers of meaning to celebrate our time? What does the locality already offer, what is here? What do the place names mean, how many of them refer to local practice, trees and woods and other clues to wild nature and our intervetition in the past? What do the field names, names of the woodlands and streams convey? Do you only find cuckoo pounds in Dorset? And what do the new names describe? Where is the history in the landscape (not the past, but history here now), where are the plants and animals particular to this place?

Taking succession seriously demands that our new forests must be created as much by natural regeneration as by planting. Let us see what wants to grow. Let us call it research in the face of global warming if that justification satisfies the fainthearted, but let us learn from nature rather than trying to teach her our short term tricks. Surely we are beyond hypothesis .... it is cheaper, more place sensitive. The challenge is to make the idea exciting to the unimaginative who only see planting by numbers as achievement.

Encouraging natural regeneration outward from existing woodland, from linear hedgerows, from railway lines, road verges and people's gardens is not a passive pursuit. Browsing animals at least will need to be kept at bay. This implies fences, which also implies gates and stiles. These can all become assets, as visual indicators bearing the marks of locality and passing them on. There may be local patterns of walling or particular ways of laying hedges, there may even be interesting fencing. If not why not work towards the creation of a local pattern? Something which leans on and reinforces the history and the wood use of the place. Stiles and gates too could form part of a simple, sensitive and silent way of letting people know where they are. Extending at whatever pace makes sense in the circumstances, made from local matenals and increasing the demand for them. Made by local people and increasing their involvement and commitment, artists and craftspeople (not necessarily local) can form part of a chain of creativity, helping to reveal, to design, if not make, things which have a richness and meaning born of understanding what is there. A thousand new ideas and as many old ways need to be explored to find ways of passing on interesting knowledge. Postcards, poetry, processions, pruning practice, performance, pollarding, picnics, apprenticeships, apple bobbing. There should be no need for 'interpretation boards', the trees should tell their own stories. Rather than cluttering places with notice boards which are all the same size and design whatever the place and the tale to tell, which reduces rather than enriches both the place and the people. Let's have field names carved in gateposts, seats which gently face the sunset, gates which beckon and give a sense of entering a new dimension.

Let us have clue after clue about the seasons. Blossom and fruit punctuate the year for us, and nowhere more dramatically and beautifully than through our wild cherry, apple, hawthorn, elder, blackthorn/sloe. Along the roadside and the path, in profusion or in single splendour in the depths of a wood. With care and understanding about the idiosyncrasies of their distribution, we can give other dimensions to the forest. Fruit trees are trees, orchards are groves. Wild and domesticated fruit trees have much to offer to the 21 century forest in town and country.

Orchards have much to offer economically, in wild life and as part of local culture. Community orchards, school orchards, city orchards could offer us pause as well as promise. Old festivals and new, particular like Arbor Day (29 May in Aston on Clun) or countrywide like Apple Day (21 October), are part of the identity of place. With Tree Dressing Day we can draw on traditions from all over the world, to draw in people of many cultures to celebrate the trees they normally take for granted. Seeking permission to perform, to light or to hang things on trees in the public domain during the first weekend in December, demands conversations with tree officers, highway department and police, suddenly bridges are crossed without anxiety, people and institutions are talking and planning together. Overstretched tree officers may begin to share responsibility with the people who 'own' the trees through familiarity. 'Those who arrive at Thekla can see little of the city, beyond the plank fences, the sackcloth screens, the scaffolding ... the wooden catwalks hanging from ropes or supported by the saw horses, the ladders, the trestles. If you ask, 'Why is Thekla's construction taking such a long time?' ..... they answer, 'So that its destruction cannot begin.'

.... 'What meaning does your construction have? ..... What is the plan you are following, the blueprint? They answer 'We will show it to you as soon as the working day is over; we cannot interrupt our work now.' Work stops at sunset. Darkness falls over the building site. The sky is filled with stars. 'There is the blueprint,' they say.' (Italo Calvino, Invisible Cities)

Imagination is crucially needed in the making of our new forests, town or country, for planting trees which seems to be born of goodness, can destroy a place. The creation of new forests has extraordinary potential to raise our sights to help us to reformulate our relations with nature, individually and socially. Our aspiration should be towards something simple and profound. The arts have much to offer. Peter Handke argued 'the duty and potential of all art as the achievement of a glimpse of humanity in the bestiality of human affairs.' Proust illuminates with a description of the poet .....

'(the poet) ... remains before the tree, and tries to close his ears to the sounds from outside and to feel once more what he had felt a moment earlier, when .... the tree had appeared before him, bearing so many white flowers on the tips of its branches that they seemed like innumerable small pellets of snow, as after a thaw. He remains before the tree, but what he seeks is no doubt beyond the tree, for he can no longer feel what he had felt, then all of a sudden he feels it again, but cannot fathom it, can go no further .... it does not seem natural that a poet should remain for an hour before this tree, gazing at how, the spring being imminent, the sure yet unconscious architectural thought known as the double cherry tree genus has arranged these innumerable small fluted white balls which until such time as they wither, will shed a faint perfume through the dark multiplicity of the tree's branches.

The poet gazes and seems to be gazing into himself and into the double cherry tree ..... (Proust, Poetry, or the mysterious laws, letter 1890s) Poetry, like succession and meaning, is about layer upon layer of richness ... literal, implied, concealed, ambiguous, mysterious, paradoxical, dynamic and demanding yet capable of revelation.

We need enchantment, from the most profound moment of realisation that we do not know the magic of the leaf, to the absorbing joy of a tree house. But just as the forestry, recreation and farming professionals need to work with meaning, so do artists. We should demand the best from them, work about and for places and the people who live and spend time there. David Nash with his trees planted with stones, his pruned ash dome and serpentine sycamores. Andy Goldsworthy with his entrance to Hooke Park and Alain Ayers oak leaves for Masham Parish Council both as part of the Common Ground New Milestones project. Jim Partridge with his fence for Grizedale Forest, and his seats for the Woodland Trust with Common Ground. Jamie MacCullough and his exceptional Beginners Way for the Exmoor Forest. Not a sculpture trail, nothing superficial or contrived, but an adventure which touched people because of the soul searching and honesty which it revealed. This was not about using the forest as a gallery, but the artist's revelation over 18 months hard work, And what of performance, festival, dance, storytelling, poetry.

We must move on from the alienating timber-by-numbers forest, but we do not have to be trapped in some Capability Brown notion of parkland for the leisured classes. We need mixed working woods, gutsy places with real things going on. But that does not mean that the best of the arts cannot be there too. Vitality and authenticity are fired by working, needing, adventuring, imagining, not by the superficial search for fun, for filling in the hours. People need to see the real and to be drawn to reverie. Adding to the depth of significance of the place is a far bigger challenge than planting new forests simply for financial gain. There is so much at stake. We must recognise that it is not an easy matter of economics or ecology, we have even harder tasks to perform which have to do with enchantment, with culture, with reclamation of knowledge of the GOOD, the WILD, the SACRED (Gary Snyder) .....

We have to recapture and progress our long understanding of trees and of the land, to reenfranchise city people and those marooned in over-farmed countryside, to nourishing the imagination. All the economic and ecological knowledge in the world will be of little use to us, if we have no imagination and humanity on which to draw, if we have lost wisdom. 7Common Ground is working with and through the power of ideas, exploring philosophies which intertwine interests in nature, culture and place with poetry. The arts can help us turn our complacent view of ordinary life upside down, can help to bring us back from abstraction, help us to reinvent, to reinterpret and to involve. We must draw in folk who have not the slightest interest in ecologically sustainable futures, who are overwhelmed by the global, bowed by the burden of history, feeling disempowered and disenfranchised by the experts. But starting small and local may help to build what democracy could and should be about. We need to create a culture of wanting to care. Only then shall we know sustainability.

Succession and poetry offer some echoes of each other, the accumulated insights of nature and culture have led both towards richness and complexity, significant layers to be read. We could stop our fight against succession, welcome the trees that want to grow amongst us, plant others of our choosing. Trees at every turn, dreamed into existence by nature and guided by institutions woi'king with local people could help us to reinvent meaning in the places close to us and bring richness back into our everyday lives. We could stop our fight against poetry and recognize that the best of humanity is born of deeper feelings.

Bibliography

Anderson, J.T.: *Farm Forestry Practices for the Students of Vocational Agriculture*, Asiatic Pub, Delhi, 2008.

Anthony B.: *Applied Ethnobotany: People, wild Plant use, and Conservation*. London, Earthscan, 2001.

Babweteera, F.: *Influence of Gap-size and Age on the Diversity and Abundance of Climbers in Budongo Forest Reserve*, Uganda. M. Sc., Makerere University, 1998.

Bagyaraj, D. Joseph: *Microbial Biotechnology for Sustainable Agriculture Horticulture and Forestry*, New India Publishing Agency, Delhi, 2011.

Bahati, J.: *Impact of Arboricidal Treatments on the Natural Regeneration of Species Composition in Budongo Forest*, Uganda. M. Sc., Makerere University, 1995.

Barrows, M.: *A Survey of the Intestinal Parasites of the Primates in Budongo Forest*, Uganda. Glasgow University, 1996.

Bor, N. L.: *A Manual of Indian Forest Botany*, Asiatic Pub, Delhi, 2010.

Chakrabarty, Falguni: *Adaptation of the Santals to the Hill-Forest Environment*, APH, Delhi, 2012.

Chakrabarty, Falguni: *Adaptation of the Santals to the Hill-Forest Environment*, APH, Delhi, 2012.

Connor Ogorzaly: *Economic Botany: Plants in our World*. Boston, McGraw-Hill, 2001.

Falguni, C: *Adaptation of the Santals to the Hill-Forest Environment*, APH, Delhi, 2012.

Gilbert Wooding Robinson: *Soils : Their Origin Constitution and Classification*, Biotech Books, Delhi, 2005.

Gupta, O.P. : *Water in Relation to Soils and Plants : With Special Reference to Agriculture*, Agrobios, Delhi, 2002.

Jackie C. Wooten *: Ethnomedicine and Drug Discovery*, Amsterdam, New York, Elsevier, 2002.

Jha, L K: *Advances in Agroforestry*, APH, Delhi, 2009.

Khare, C. P. *Encyclopedia of Indian Medicinal Plants: Rational* International Book Distributing Co., Delhi, 2008.

Kotwal, P.C. and Sujoy Banerjee *: Biodiversity Conservation in Managed Forests and Protected Areas*, Agrobios, Delhi, 2004

Kundu, S.S; O.P. Chaturvedi, J.C.Dagar and S.K. Sirohi*: Environment, Agroforestry and Livestock Management,*

Macself, A.J. : *Soils and Fertilizers*, Satish Serial Pub, Delhi, 2005.

Mani, A K; R Santhi and K M Sellamuthu: *Fundamentals of Forest Soils*, Satish Serial Pub, Delhi, 2008.

Miller, P. L.: *Some Dragonflies of the Budongo Forest*, Western Uganda. Opusc, 1993.

Sathe, T.V.: *A Textbook of Forest Entomology*, Daya, Delhi, 2009.

Sett, Rupnarayan*: Environmental Science : Botanical and Forestry Perspective*, Narendra Pub, Delhi, 2012.

Sharma, Ajay: *A Text Book of Forest Trees Entomology*, International Book Dist, Delhi, 2007.

Sharma, R.C. and J.N. Sharma *: Challenging Problems in Horticultural and Forest Pathology*, Indus, Delhi, 2005.

Shiel, D.: *The Ecology of Long-term Change in a Ugandan Rainforest*. D. Phil., Oxford University, 1997.

Shivanand Tolanur: *Practical Soil Science and Agricultural Chemistry*, International Book Distributing, Delhi, 2004.

Thiericke, R. *: Drug Discovery from Nature*. Berlin, New York, Springer, 1999.

Trivedi, P.C. : *Plant Physiology in Agriculture and Forestry*, Aavishkar Pub, Delhi, 2009.

Vyas, G P D: *Community Forestry*, Agrobios, Delhi, 2006.

Ward, H. Marshall: *Trees a Handbook of Forest-Botany*, Asiatic, Delhi, 2001.

Western Therapy, Ayurvedic and other Traditional Usage, Berlin, New York, Springer, 2004.

Wickens, G. E.: *Economic Botany: Principles and Practices*. Dordrecht, Boston, Kluwer Academic, 2001.

Index

H

I

J

M

N

O

P

R

S

T

U

V